Using Open Data to Detect Organized Crime Threats

Henrik Legind Larsen · José María Blanco
Raquel Pastor Pastor · Ronald R. Yager
Editors

Using Open Data to Detect Organized Crime Threats

Factors Driving Future Crime

Springer

Editors
Henrik Legind Larsen
Department of Electronic Systems
Aalborg University
Esbjerg
Denmark

José María Blanco
Spanish Law Enforcement Agency
Guardia Civil
Madrid
Spain

Raquel Pastor Pastor
Isdefe S.A.
Madrid
Spain

Ronald R. Yager
Machine Intelligence Institute
Iona College
New Rochelle, NY
USA

ISBN 978-3-319-52702-4   ISBN 978-3-319-52703-1   (eBook)
DOI 10.1007/978-3-319-52703-1

Library of Congress Control Number: 2017930272

© Springer International Publishing AG 2017
This work is subject to copyright. All rights are reserved by the Publisher, whether the whole or part of the material is concerned, specifically the rights of translation, reprinting, reuse of illustrations, recitation, broadcasting, reproduction on microfilms or in any other physical way, and transmission or information storage and retrieval, electronic adaptation, computer software, or by similar or dissimilar methodology now known or hereafter developed.
The use of general descriptive names, registered names, trademarks, service marks, etc. in this publication does not imply, even in the absence of a specific statement, that such names are exempt from the relevant protective laws and regulations and therefore free for general use.
The publisher, the authors and the editors are safe to assume that the advice and information in this book are believed to be true and accurate at the date of publication. Neither the publisher nor the authors or the editors give a warranty, express or implied, with respect to the material contained herein or for any errors or omissions that may have been made. The publisher remains neutral with regard to jurisdictional claims in published maps and institutional affiliations.

Printed on acid-free paper

This Springer imprint is published by Springer Nature
The registered company is Springer International Publishing AG
The registered company address is: Gewerbestrasse 11, 6330 Cham, Switzerland

Foreword

Terrorism and serious and organized crime remain among the threats of greatest concern in modern society. Regardless of whether they are motivated by profit or extremism, criminal and terrorist networks are extremely flexible and have repeatedly shown their ability to evolve rapidly in response to environmental changes. Increasing mobility and the tremendous opportunities offered by technology are contributing to this rapid adaptation.

Law enforcement agencies are therefore facing new threats which are emerging at an accelerated pace. Increasingly, these threats are not contained within a single jurisdiction, and the public funds available to tackle them have, in many cases, been reduced due to the economic downturn of the last few years. The only possible answer is for law enforcement to complement traditional methods by exploiting all opportunities offered by new technologies and the ever-growing amount of open source information available. Clearly, gaining access to all possible sources of information is not enough; it must also be possible to analyze and derive meaning from increasingly large and diverse data sets.

In cooperation with law enforcement agencies throughout Europe, the partners of the ePOOLICE consortium developed the prototype of an environmental scanning system to explore open source information in order to detect new emerging terrorist and serious and organized crime threats. The idea was to combine state-of-the-art technologies and concepts (including statistical and text data analysis) with social science modeling of the appearance and evolution of crime phenomena. Identifying macro-environmental factors influencing the emergence of crime, as well as defining the degree of their impact, was one of the most challenging aspects of the project. The ePOOLICE initiative was a remarkable opportunity to intensify the cooperation between law enforcement, academia, and the private sector. The most promising new technologies in the fields of information management and data fusion were applied to develop a system of tools for the gathering of information via the Internet, turning it into actionable intelligence and providing alerts on emerging threats. The law enforcement partners contributed use cases, sharing their experience about how crime evolves, how criminals target their victims, and how they organize their activities. The EU Commission acknowledged the system to be a successful proof of the concept.

I hope that the results of the project will lead to a functional software solution which our analysts can use in the future to support law enforcement decision making. More broadly, cross-border and cross-sector partnerships and innovative approaches such as those described in this book will be crucial to the ability of law enforcement agencies to prevent and combat crime in the digital age.

The Hague, Netherlands   Rob Wainwright
                                                                           Director of Europol

Preface

In countering serious and organized crime, including terrorism, foresight of emerging threats is extremely important as it provides a strategic advantage over the potential adversaries, allowing the security and law enforcement forces to organize their capacity as needed to meet the threats. The emergence of such threats is to a high degree driven by the development of a combination of certain factors in the environment of the PESTLE (political, economic, social and demographic, technological, legal, and environmental) situation in a region.

Open data provide a rich basis for extracting information and deriving strategic intelligence, allowing us to spot development of such crime-relevant factors early and thereby to foresee threats that are likely to emerge. For our purpose, the open data comprise publicly available data and information in any form, in particular electronically through the Internet, which can be accessed legally and ethically.

Due to the huge amount of open source information, in various forms and of various qualities, there is a need for a system solution for monitoring the sources and extracting valid relevant information in due time. The project ePOOLICE (www.epoolice.eu), is a project in the 7th Framework Programme of the European Union, aimed at developing such a solution.

This book contains a collection of papers on a topic considered particularly important for this purpose, namely "factors driving future crime," which was the topic of one of the ePOOLICE workshops held as discussion fora among technicians, law enforcement agencies, and other stakeholders. The aim of the workshop was to highlight and review some factors considered as catalyzer of organized crime. The objective was to evaluate the trend of the selected factors, identify the potential weak signals, and suggest the appropriate response (e.g., combination of technological resources and human actors that serves to improve the process of assessing new organized crime threats).

This book comprises both relevant theory (strategic analysis, measures of crime, methodology, etc.) and presentation of practical cases with focus on relevant aspects of these cases. It describes several interesting results and ideas that were presented and discussed during the workshop. Several issues and questions provide challenges for future work. This book is organized as follows:

Chapters 1–2 Introduce Some Key Concepts

Chapter "Organised Crime as a Framework Concept" provides an overview of the situation of the research of organized crime in its different dimensions. It provides a discussion of organized crime as a reference framework, in order to delimit the scope of this crime as complex criminological phenomena, and then presents and analyzes some key issues in organized crime research.

Chapter "Measuring Organised Crime: Complexities of the Quantitative and Factorial Analysis" focusses on the measurement of organized crime and the need for a quantitative approach; it further proposes and discusses a methodological approach to quantitative data and the development of trustworthy indicators.

Chapters 3–9 Focus Methodological Aspects

Chapter "Scanning of Open Data for Detection of Emerging Organized Crime Threats—The ePOOLICE Project" presents the ePOOLICE project that was aimed at developing efficient and effective system for scanning the open data detection emerging organized crime threats. It introduces the problem and the methodological framework considered and outlines the principles of the solution developed and how the project was organized.

Chapter "Foresight and the Future of Crime: Advancing Environmental Scanning Approaches" introduces and provides a detailed outline of environmental scanning and the early warning systems' framework for crime detection and prevention in a predictive policing context using foresight methodology such as weak signals, facilitating factors, horizon scanning, and leading indicators.

Chapter "Big Data—Fighting Organized Crime Threats While Preserving Privacy" deals with the challenge of balancing privacy and national security in connection with big data-driven sense-making systems for crime fighting. Throughout, it refers to the prototype system from the ePOOLICE project to illustrate the various issues discussed.

Chapter "Horizon Scanning for Law Enforcement Agencies: Identifying Factors Driving the Future of Organized Crime" proposes and articulates a methodology for law enforcement agencies to identify factors driving future crime. For this purpose, it outlines the use of horizon scanning for Open Sources of Information (OSINF), i.e., open data, leading to the development of actionable Open Sources of Intelligence (OSINT), which is fast becoming an integral part of the prevention, investigation, and detection of contemporary organized crime.

Chapter "Macro-environmental Factors Driving Organised Crime" proposes a new methodological model, as an intelligence process, for analyzing the future of organized crime in the times of VUCA (volatility, uncertainty, complexity, and ambiguity) environments. The model provides a schematic view on trends and security outcomes for different PESTLE dimensions. For each trend, the framework applied shows drivers, indicators, degree of evidence, time range and outcomes, qualified by the degree of uncertainty. This model allows monitoring future development of identified trends, generating knowledge and intelligence to support decision making.

Chapter "Extracting Future Crime Indicators from Social Media" introduces a generic intelligence-driven process for extracting indicators from social media, describes a concrete implementation, and illustrates its interest and strength on two experiments. Based on the experimental findings, it proposes a generic framework for the construction of crime indicators from social media feeds.

Chapter "Organised Crime, Wild Cards and Dystopias" considers the challenges for foresight of future organized crime threats and presents a methodological approach, including techniques and tools that can help to create this vision of the future from the future (and not from the present, as proposed in previous chapters): wild cards, scenarios, backcasting, "what if," and dystopias. Imagination and creativity must be a part of the sources of knowledge and the methodological approaches are needed to understand and manage the future.

Chapters 10–12 Present and Analyze Some Concrete Use Cases

Chapter "Operation Golf: A Human Trafficking Case Study" presents a UK and Romania Joint Investigation, Operation Golf, tackling Romanian organized crime and child trafficking. It provides detailed information about how the investigation was organized and run, as well as about the outcome of the investigation. In a prologue by the editors, it is outlined how open data intelligence can help such investigations in several ways.

Chapter "Radicalization in a Regional Context: Future Perspective on Possible Terrorist Threats and Radicalization" deals with several theories of radicalization and the extremist ideologies behind. It presents a number of possible future radicalization scenarios and highlights drivers and indicators. It also focuses on the role of the media to illustrate the use and abuse of the media by violent extremist groups.

Chapter "Why Do Links Between Terrorism and Crime Increase?" examines the nexus terror crime and the various insights in the field, its possible forms, and the main criminal activities involved. The factors and enhancers of the current convergence between terrorism and crime are analyzed. Special attention is given to the causes that enable and prompt terrorist involvement in criminal acts and relationships with criminal groups.

Esbjerg, Denmark	Henrik Legind Larsen
Madrid, Spain	José María Blanco
Madrid, Spain	Raquel Pastor Pastor
New Rochelle, NY, USA	Ronald R. Yager

Contents

Part I Key Concepts

Organised Crime as a Framework Concept 3
Andrea Gimenez-Salinas Framis

**Measuring Organised Crime: Complexities of the Quantitative
and Factorial Analysis** 25
Daniel Sansó-Rubert Pascual

Part II Methodology

**Scanning of Open Data for Detection of Emerging Organized
Crime Threats—The ePOOLICE Project** 47
Raquel Pastor Pastor and Henrik Legind Larsen

**Foresight and the Future of Crime: Advancing Environmental
Scanning Approaches** 73
Martin Kruse and Adam D.M. Svendsen

**Big Data—Fighting Organized Crime Threats While Preserving
Privacy** 103
Anne Gerdes

**Horizon Scanning for Law Enforcement Agencies:
Identifying Factors Driving the Future of Organized Crime** 119
Timothy Ingle and Andrew Staniforth

Macro-environmental Factors Driving Organised Crime 137
José María Blanco and Jéssica Cohen

Extracting Future Crime Indicators from Social Media 167
Thomas Delavallade, Pierre Bertrand and Vincent Thouvenot

Organised Crime, Wild Cards and Dystopias 199
José María Blanco and Jéssica Cohen

Part III Use-Cases

Operation Golf: A Human Trafficking Case Study 223
Bernie Gravett

Radicalization in a Regional Context: Future Perspective on Possible Terrorist Threats and Radicalization 241
Holger Nitsch

Why Do Links Between Terrorism and Crime Increase? 261
Luis de la Corte Ibáñez and Hristina Hristova Gergova

Index 281

Editors and Contributors

About the Editors

Henrik Legind Larsen, professor, Aalborg University, Denmark, and CEO of Legind Technologies, Denmark, was scientific-technical manager of the ePOOLICE project and organizer of the workshop on Factors Driving Future Crime that gave rise to this book. His main research is in intelligent information systems, flexible information access, and fuzzy logic-based technologies. He is co-founder of the international conferences series on Flexible Query Answering Systems (FQAS) and served as general chair of several conferences in this series. He is member of the editorial boards of the journals Fuzzy Sets and System and International Journal of Uncertainty, Fuzziness, and Knowledge-based Systems. His scientific publication comprises 78 papers.

José María Blanco is the head of the Centre of Analysis and Foresight in Guardia Civil, co-director of Strategic Studies and Intelligence in the Institute of Forensic Sciences and Security (University Autónoma, Madrid), adviser in the Universitary Institute of Homeland Security, editor of the academic–professional journal "Cuadernos de la Guardia Civil," professor of intelligence studies, terrorism, and organized crime in several universities, researcher in European Union funded projects, and editor of the book "Seguridad Nacional: amenazas y respuestas," ed. LID (2015). He has published several papers, articles, and reports in media, think tanks, and academic journals.

Raquel Pastor Pastor received her M.Sc. degree in telecommunications engineering from Polytechnic University of Madrid. She has a professional career of over 25 years, having worked both nationally and internationally at private and public sectors. She joined Isdefe, Spanish state-owned company of Defence Ministry, in 2003 as responsible for the Software Engineering Group within the Defence and Security Directorate, and worked on several intelligence and electronic warfare projects for the Spanish Defence Ministry. As contract manager of FP7 European OSINT projects within the Surveillance Technologies Department of Isdefe, she worked as WP leader in European FP7 co-funded VIRTUOSO project on OSINT and Intelligent surveillance and enhancing border security and she managed as project coordinator the European FP7 ePOOLICE project, on Early Warning and Environmental Scanning for the detection of organized crime threats.

Ronald R. Yager is director of the Machine Intelligence Institute and Professor of Information Systems at Iona College. He is editor and chief of the International Journal of Intelligent Systems. He has published over 500 papers and edited over 30 books in areas related to fuzzy sets, human behavioral modeling, decision making under uncertainty, and the fusion of information. He is among the world's most highly cited researchers with over 57,000 citations in Google Scholar.

He was the recipient of the IEEE Computational Intelligence Society Pioneer award in Fuzzy Systems. He received the special honorary medal of the 50th Anniversary of the Polish Academy of Sciences. He received the Lifetime Outstanding Achievement Award from International the Fuzzy Systems Association. He recently received honorary doctorate degrees, honoris causa, from the Azerbaijan Technical University and the State University of Information Technologies, Sofia Bulgaria. Dr. Yager is a fellow of the IEEE, the New York Academy of Sciences, and the Fuzzy Systems Association. He has served at the National Science Foundation as program director in the Information Sciences program. He was a NASA/Stanford visiting fellow and a research associate at the University of California, Berkeley. He has been a lecturer at NATO Advanced Study Institutes. He was a program director at the National Science Foundation. He was a visiting distinguished scientist at King Saud University, Riyadh Saudi Arabia. He was an adjunct professor at Aalborg University in Denmark. He received his undergraduate degree from the City College of New York and his Ph.D. from the Polytechnic Institute New York University. He is the 2016 recipient of the IEEE Frank Rosenblatt Award, the most prestigious honor given out by the IEEE Computational Intelligent Society. He recently edited a volume entitled Intelligent Methods for Cyber Warfare.

Contributors

Pierre Bertrand Thales Communication and Security, Gennevilliers, France

José María Blanco Centre of Analysis and Foresight, Guardia Civil, Madrid, Spain

Jéssica Cohen International Security Intelligence Analyst, Private Sector, Madrid, Spain

Luis de la Corte Ibáñez Universidad Autónoma de Madrid, Madrid, Spain

Thomas Delavallade Thales Communication and Security, Gennevilliers, France

Andrea Gimenez-Salinas Framis Universidad Autónoma of Madrid, Madrid, Spain

Anne Gerdes Department of Design and Communication, University of Southern Denmark, Kolding, Denmark

Bernie Gravett Specialist Policing Consultancy Ltd., Oxfordshire, UK

Hristina Hristova Gergova Private Sector, Madrid, Spain

Timothy Ingle West Yorkshire Police, Wakefield, England, UK

Martin Kruse Copenhagen Institute for Futures Studies, Copenhagen, Denmark

Henrik Legind Larsen Department of Electronic Systems, Aalborg University, Aalborg, Denmark

Holger Nitsch Department for Policing, College for Public Administration in Bavaria, Bavaria, Germany

Daniel Sansó-Rubert Pascual Security Studies Center (CESEG), University of Santiago de Compostela, Santiago de Compostela, Spain

Raquel Pastor Pastor ISDEFE, Madrid, Spain

Andrew Staniforth West Yorkshire Police, Wakefield, England, UK

Adam D.M. Svendsen Copenhagen Institute for Futures Studies, Copenhagen, Denmark

Vincent Thouvenot Thales Communication and Security, Gennevilliers, France

Part I
Key Concepts

Organised Crime as a Framework Concept

Andrea Gimenez-Salinas Framis

1 Introduction

The subject of organised crime has traditionally attracted the attention of experts and laypersons. Evidence of this is the vast number of series and films on the television and in theatres on the subject of mafia organisations and the mystery surrounding criminal organisations. However, it is a particularly difficult field from an academic perspective, and very limited results have been accomplished in comparison with other areas of crime, such as juvenile delinquency or common crime, which is understood as individual crime against sexual freedom, physical integrity or property (Gimenez-Salinas et al. 2009; De la Corte and Gimenez-Salinas 2010).

The study of organised crime dates back to the 1920s when Frederic Trasher (1927) and Ladesco (1929) first started researching the subject. However, in the early decades, research was generally more descriptive and offered scarce empirical results that could be used to support theoretical constructions (von Lampe 2011). Over the years, there was a period during which the focus was primarily on defining this phenomenon and finding a valid concept academically. Hagan (2006) provided a good summary of the different definitions representing a wide range of concepts which debate was finally settled by the United Nations Convention against Transnational Organised Crime in 2000. By contrast, in the last decades there have been more empirical studies in the field of organised crime, although the results are still a long way from reaching the theoretical conclusions of other areas of crime. The reasons for this are threefold:

(a) This is a new area, in comparison with others which consensus on their distinctiveness and conceptual definition dates back to 2000 if we consider this date as a milestone in terms of international consensus regarding the definition of

---

A.G.-S. Framis (✉)
Universidad Autónoma of Madrid, Madrid, Spain
e-mail: andrea.gimenezsalinas@inv.uam.es

© Springer International Publishing AG 2017
H.L. Larsen et al. (eds.), *Using Open Data to Detect Organized Crime Threats*, DOI 10.1007/978-3-319-52703-1_1

organised crime (by the United Nations Convention). (b) The empirical investigation of these organisations is especially complicated and sometimes dangerous, and the risks involved limit the possibility of studying and accessing samples of subjects and organisations which can be used to further develop our knowledge in these fields, respectively. (c) Finally, the results of the research to date are just theoretical developments that enforcement officers or politicians find difficult to apply when taking decisions on the matter, and thus, future investigations should focus on bridging the gap between research and practice.

We now provide a brief overview of the situation of the research of organised crime in its different dimensions. Firstly, we provide a definition as a reference framework and to delimit the scope of this matter within the currently existing criminological phenomena. In addition, we analyse some other lines of investigation of organised crime: how non-conventional crimes is measured (van Dijk 2007), which current tools are far from perfect to calibrate the dimensions of a complex phenomenon; the genesis of this phenomena and the criminogenic creation and influence of criminal organisations; and the progress in terms of an individual explanation for a type of crime, which is very different to common crime because it requires a higher degree of qualification, planning, expertise and international contacts.

2 Origin and Meanings of the Concept of Organised Crime

Since the term organised crime first appeared in the USA in the 1920s, the concept has significantly changed because of how it is perceived by society and by the authorities in charge of fighting it. In these 90 years, there has been a first discovery stage, then a stage of local concern about foreign organisations setting up on American soil and a last stage, in which organised crime became an international concept with multiple ethnic origins. In Europe, however, more importance has been given to the economic approach of organised crime integrated in the market economy. An examination of these phases is carefully based on the description provided by von Lampe (2001), who carried out an interesting quantitative and qualitative study of the concept over time and in the USA, using media and documentary sources.

The notion of organised crime was first used at the Chicago Crime Commission, which was created in 1919 by bankers, businesspersons and lawyers to promote legislative reform on the subject of crime (von Lampe 2002). This concept was coined in a much wider sense than its current meaning, due to inexperience and its novelty then. The initial concern centred on the emergence of a "professional criminality", formed by a large number criminals that profited greatly from crimes to the amazement of the general public, while the authorities took no action. These criminals would take advantage of the existing power vacuums to profit from criminal activity such as crimes against property.

Ten years later, the real threat of this phenomena began to emerge and any prior positive connotations began to fade. This is when talk of racketeering or crime syndicates first began, which through business associations and trade unions tried to control the market and bribed businesspersons (Lashly 1930). Thus, further steps were taken in terms of repression by demanding sterner action and greater involvement by law enforcement. After this stage, the debate went quiet for several years until it picked up again intensely in the 1950s when the Kefauver Senate Committee was set up to investigate inter-state commerce by organised crime in the USA. The conclusions of this Committee in 1951 had a major impact because they revealed that numerous organised crime groups had come together to create a "sinister criminal organisation known as the Mafia, whose tentacles are found in many large cities... Its leaders are usually found in control of the most lucrative rackets" (US Senate 1951). Consequently, organised crime ceased to be a local problem in just one state and became part of the Italian-American Mafia. On the other hand, the committee's conclusions came at a time when a national meeting of Italian-American members of the Mafia was discovered to have taken place near Apalachin in the State of New York. In the same way, the McClellan Committee set up in 1960 gained more knowledge of the Cosa Nostra through the testimony of a member of the organisation, Joseph Valachi, who testified against some important leaders and revealed the internal workings of the organisation.

These revelations were given a scientific systematisation by Donald Cressey in a book titled *The theft of the Nation* (1979), whose decision to include alien conspiracy in the notion of organised crime was essential (Paoli 2002) even if it was much disputed by most American social scientists who believed it was ideological, politically motivated and lacking empirical rigour (Smith 1976; Moore 1974; Hawkins 1969). In response to this rejection, these authors opted for other approaches such as the notion of "illicit enterprise", assimilating organised crime to the business surroundings. From that moment, some authors began to explore the most economic aspect of this concept, focusing especially on the illicit activity, which generated a financial benefit: the provision of goods and illicit services over nationalities and cultural aspects (Smith 1976; Block and Chambliss 1981; Hagan 1983).

The economic perception of organised crime also inspired the studies on organised crime in Europe in the 1970s and 80s by authors such as Arlacchi (1986) or Smith (1980), who argued that modern members of the Mafia had transformed their strategies and functions along with economic modernisation, making their activity more entrepreneurial than intermediary or based on protection. Other authors followed this economic approach yet disagreed with Arlacchi, because they believed that these functions were present from the early days of the Mafia (Catanzaro 1991).

Both aspects are significant in organised crime, be it the persons who form the group or the illicit criminal activities carried out to ultimately obtain a financial benefit. This is why both concepts were ultimately considered essential for the notion of organised crime, even though in the 1990s comparing organised crime with legitimate companies prevailed over the notion of mafia, which was no longer associated with the Italian model, but rather defined any criminal organisation irrespective of the nationality of its members (Paoli 2002). The concept of organised crime was not free of controversy after the 1990s. From the 1990s to the 2000s, the core of the scientific debate on this phenomena centred on a definition that could encompass all the aspects and dimensions of organised crime. We now explain the main dimensions of criminal behaviour, the differences with other related group phenomena and the agreed and established definitions to date.

3 Concept and Dimensions of Organised Crime

The type of criminal activity analysed in this chapter falls into a category that criminology calls *complex or non-conventional crime* (von Lampe 2004; Vander Beker 2004; Albanese 2000, 2008). These are different to conventional offences, which are mainly unlawful acts committed by a single perpetrator, because of the diversity of dimensions and perpetrators behind the criminal act. These multiple dimensions not only make defining this particular phenomena more difficult, but also directly affect its scientific assessment, description and approach (Gimenez-Salinas et al. 2009; van Dijk 2007). In this regard, organised crime takes its name from the fact that one of its essential conditions is that it is a group of offenders with diverse members and varied complexity, which main activity is the continuous provision of generally illegal goods or services (illicit markets) and which carry out their actions or activities through corruption, violence, penetration of legitimate businesses and money laundering (De la Corte and Gimenez-Salinas 2010; Gimenez-Salinas et al. 2009). Thus, organised crime comprises three dimensions, each of which has its own components and dynamics, which need to be understood in a global context.

(a) The first dimension comprises groups or organisations of organised crime, which exist in multiple forms depending on how many members they have, their structure, their internal hierarchy, members' division of labour and their local or international presence.

(b) The second dimension comprises illegal markets or the provision of illegal goods or services, which the organisations develop as their main activity to generate profits. These illegal markets can exist in different forms and can be developed in different ways depending on the product being traded (drugs, weapons, alcohol, tobacco, protected species, immigrants, organs, diamonds, among many others). As any other market, their existence is the result of an imbalance between a limited legitimate offer and its corresponding demand. This disparity can be caused by different circumstances: the absence of or an insufficient legitimate offer, because it is subject to regulation, restriction or control (weapons, immigration etc.) or due to the product's higher price because it is subject to taxes or levies (e.g. tobacco or alcohol) (De la Corte and Gimenez-Salinas 2010; von Lampe 2016).

(c) The third dimension comprises activities that, in an instrumental way, organisations generally carry out to remain underground in a hostile environment subject to constant surveillance. The most remarkable feature of these instrumental activities is that they are generally found transversally in all organisations with variable intensity depending on the capacity and size of the organisation and the benefits that its local environment offers. Four activities are identified:

- *Corruption* as a means of attaining areas of immunity by bribing police agents, customs officers or anyone who has control over or any way of benefiting the illegal business, or by paying public officers to allow the organisation to develop its business (civil servants who knowingly grant false residence permits to human traffickers or licences or concessions to criminal organisations, etc.).
- *Violence* is another method used by organisations, which carry out their activity in an environment that is regulated by a legal system to which they cannot resort to resolve their problems or disputes. Violence is a method that, for organised crime, has various uses, such as resolving external disputes—setting limits for competing organisations—enforcing internal discipline or developing the business itself when this requires an element of violence, threats and coercion. With regard to the latter, we refer, for instance, to illegal businesses such as human trafficking for sexual exploitation or child trafficking.
- *Penetrating* or *using the legal economy* is a strategy that organisations often use to conceal their illegal activities, launder illegally earned money, shift stolen or unlawfully obtained goods. This includes, for example, car dealers which are used to sell stolen vehicles or art galleries which are used for art trafficking purposes. It is common for the legal and illegal activities of large criminal organisations to be intermingled, and this poses a serious threat to the economic sector in which they are included, since their presence and activity gives rise to unfair competition.
- *Money laundering* is a necessary process for organisations that accrue illicit benefits and need to use them for legal purposes. Therefore, reasonably large organisations require more or less sophisticated ways of concealing the illegal source of that income; from sending the money to other countries covertly and circumventing the financial system, to more complex ways such as using shell companies or tax haven jurisdictions.

By defining organised crime, the conceptual barriers between organised crime and other similar criminal behaviour can be framed and delimited, such as terrorism or juvenile gangs, which share some traits with organised crime. All three cases include a group as a unit of active subjects who participate in criminal activities. However, the difference mainly centres on the goals pursued with the illegal activities. Organised crime only seeks an economic gain, while terrorist groups have an ideological or political element, and money or wealth is only an instrumental means of achieving their aims. On the other hand, the aim of juvenile gangs is to build their own identity, and an alternative means and style of life for their members (Vazquez González and Serrano 2007).

Despite their different aims, other aspects of their means or methods could be similar, especially between organised crime and terrorist groups. We refer, for instance, to the methods of financing, the ways of implementing organisational discipline or violence as a way of sending out messages to the public or as a sign of authority and strength to society and to those in power. A clear example of similar means between terrorism and organised crime is the killings and assassinations committed by organised crime in Mexico, when Felipe Calderón was in power. These aimed to coerce and threaten society and directly lock horns with the government. On the other hand, the conceptual barriers described above are sometimes difficult to discern from reality, and some organisations can be found in between both phenomena, forming a type of hybrid category. The Colombian group, known as FARC, is a good example of this. It started off as a terrorist group but became an organised crime group due to its main involvement in drug trafficking; or the 18th Street or Mara Salvatrucha gangs, which started off as juvenile gangs but have become fully-fledged organised groups.

So far we have covered the criminological definition of organised criminal groups regarding which there is currently consensus in criminological literature (Finckenauer 2010) after many years of conceptual debate. However, when this definition is transposed to national and international legislations, consensus is not as straightforward, especially with regard to the defining characteristics of organised crime organisations.

On the one hand, some international definitions, such as that deriving from the United Nations Convention against Transnational Organized Crime (2000), state that an organised criminal group is "a structured group of three or more persons, existing for a period of time and acting in concert with the aim of committing one or more serious crimes or offences [...], in order to obtain, directly or indirectly, a financial or other material benefit". In the same vein, according to Council Framework Decision 2008/841/JHA, a criminal organisation is "a structured association, established over a period of time, of more than two persons acting in concert with a view to committing offences which are punishable by deprivation of liberty or a detention order of a maximum of at least four years or a more serious penalty, to obtain, directly or indirectly, a financial or other material benefit". These two definitions are similar, even though they do not comprise all the indicators mentioned above, such as the characteristics of criminal organisations; they only refer to the group, the continual and serious activity, and its aim. More in line with the abovementioned criminological concept are the indicators used by Europol to classify organised crime organisations. In this regard, Europol requires the coexistence of four mandatory and some other optional indicators. The four mandatory indicators are the following: the collaboration of two or more people, the quest for a means of power, continuity over time and the suspicion of participation in serious crimes. At least two of the following optional indicators are also required: distribution of specific tasks, control mechanisms and internal discipline, international activity, use of violence and intimidation, use of business and economic structures, involvement in money laundering, and political and media influence.

As shown, the concept of organised crime is far from straightforward and is made up of numerous dimensions, all of which are important for its existence and development. On the other hand, each of these dimensions is not static but rather constitutes gradual dimensions, which are consistent with the size and complexity of the criminal groups under analysis. It is not the same to talk about a network of ten members dedicated to distributing cocaine in a specific geographical area than about the Camorra or the 'Ndrangheta, which are big and influential organisations with international contacts, high capacity for corruption, and which exercise major control over the territory and the illicit markets in which they operate; they also move substantial volumes of illegal money which needs to be laundered. Its complexity and this dimensional variation obscure a very important aspect of the investigation: the volume of organised crime existing in a specific period of time or territory, which we call measurement of organised crime. We examine this subject below.

4 Measuring and Evaluating Organised Crime and their Difficulties

One of the classic areas of research in the criminological field is measuring criminal phenomena. Criminology has taken great steps to design measuring tools and indicators to approximate the volume of crimes committed under the different types of crimes. However, the traditional instruments of criminology cannot be used for this area of crime. Victimisation surveys or self-report surveys used to measure common crime are not valuable to analyse organised crime for two reasons: (a) Because it is an unconventional concept which cannot be confined to a specific act such as sex crimes or crimes against property. Organised crime groups commit multiple criminal acts, some inherent to their illegal business (selling drugs, gun trafficking, etc.) and others necessary to develop their activity (coercion, threats, murder, bribes, etc.). (b) Because its effects or damage do not always appear on individual victims, which can be used as a source of information to know more about the crime victimisation.

As a consequence, to measure organised crime, victimisation surveys are usually carried out on companies, as they have the highest level of victimisation with regard to specific acts of organised crime (extortion, bribes, fraud, etc.). The following are some examples of victimisation surveys conducted on companies used to measure some forms of organised crime: the *International Commercial Crime Survey* (ICCS 1994) carried out by UNICRI and which measures conventional crimes and some forms of corruption, the *International Crime Business Survey* (ICBS 2000) conducted by the UNODC on corruption, fraud and extortion, and the *Crime and Corruption Business Survey* (CCBS) conducted by UNODC (2007) which measures bribes, corruption, fraud, extortion and conventional crimes against companies. Finally, at a European level, the *European Business Crime Survey* (Dugato 2013) was conducted by Transcrime and Gallup Europe to measure bribes and corruption, cybercrime piracy and counterfeiting; protection money and usury.

Using official records as a reference (police sources for complaints or judicial sources for legal proceedings) to measure organised crime, gives rise to numerous difficulties, especially if we compare the figures at an international level. In Europe, following the establishment by Europol of the common indicators to classify organised crime groups, there exists comparative information which gives further details about the groups arrested and the subjects investigated and arrested for organised crime, even if this information is not public. Nevertheless, elsewhere, arrests do not accurately reflect the real extent of this phenomena, especially in countries with high levels of corruption, with significant political intervention in the fight against crime and police investigations with limited success. In these cases, there is a negative correlation between the the figures of organised crime and its actual existence in society (van Dijk 2007).

In light of these difficulties, indirect measuring has been developed to approximate the dimensions of organised crime. One way is to measure organised crime by estimating its illicit markets. Knowing the volume of drug trafficking or tobacco contraband or human trafficking allows us to measure organised crime from within the main business of criminal organisations. In this regard, there are various approximations that focus on measuring using different variables: the offer or the object of the business (e.g. when we measure how many hectares of a particular type of drug are cultivated), or demand (levels of consumption or products available in the market or number of people traded); value of a particular market (through the price of the product or the components in the different phases of the business (Reuter et al. 1990); police detection of an illegal product (through police seizures); benefits obtained from a specific market (goods confiscated from criminal groups or value of the seizures in an illegal market); and, finally, the impact or evaluation of the damage to a specific market (costs of the damage).

Another method of approximation is measuring the transversal or instrumental activities comprising the third dimension of organised crime, which can be found in most organisations to a greater or lesser extent (corruption, money laundering and violence). If these activities can be measured, and those related to organised crime can be disaggregated, there should be an indirect measure of the same. For example, indicators such as underground economy could be used to measure local money laundering; homicides or kidnappings could be used to measure violence injuries; and a corruption perception index can be used to measure corruption. Taking into account these indirect measures as a reference, there have been attempts to combine them into composite indices for an international measurement. Such is the case of the composite index for organised crime perception developed by van Dijk (2007) which, although has not been repeated annually, is the only input to designing a method to measure international organised crime. This composite index combines the following: the degree of organised crime perception according to various surveys conducted internationally on how organised crime is perceived (World Economic Forum, World Competitiveness Reports (WEF), Business Executive

Surveys, 1997–2003; Merchant International Group 2004 (2003) and BEEPS 1996; Money laundering and informal sector (World Economic Forum Business Executive Survey, 2003; High Level of corruption index in Kaufmann et al. (2003); and the Unsolved Homicides (8th UN Survey on Crime and Justice, 2002).

4.1 Risk Assessment Tools

Apart from measuring organised crime dimensions, numerous studies have developed risk assessment models. The threat of criminal groups is gradual and very diverse depending on their configuration, activities and level of economic and political penetration in society. Therefore, quantitative measuring does not suffice to identify the level of threat of criminal groups; assessment models are required, with which the various dimensions and level of threat can be adjusted. The advantage of these models is that they can be valuable in decision-making processes on prevention and repression. The following are some examples of these types of models, applied to criminal groups, illicit markets or to the structural elements that form the breeding ground for these groups and their development.

There are two main ways of evaluating markets: assessing the market and its impact on the society in which it is located and assessing the vulnerability of creating a specific market:

(a) Two examples of market assessment and its consequences are provided: the LERNA and ARKO projects, developed by the Queensland Police Service (Australia) in 2000 (Black et al. 2001). Both are based on the assumption that illicit markets can be treated as legitimate markets and, thus, that market evaluation analysis can be used to analyse their impact. This type of analysis can be carried out at a macro- or micro-level. At a macro-level, the analysis determines impact on demographic or ethnographic factors or on aspects related to health and crime. At a micro-level, analysing a specific market brings to light barriers to entry, forms of negotiation by suppliers or consumers, ploys by actors in the market, etc. Therefore, this type of risk assessment exposes vulnerabilities which can be used to steer police action better, and it also provides an insight into the specific features of each illegal market and its impact on society.
(b) Other assessment models measure the vulnerability or predict the risk (probability) of organised crime being present in the legitimate economy. To this end, aspects such as organised criminality and economic sector regulations or standards are taken into account. The work of Ruggiero (1996, 2000), Hobbes (2001), Edward and Gill (1999) and Vander Beken (2004) are particularly noteworthy in this area.

To assess organisations and measure their level of risk or threat, the contributions of the Bundeskriminalamt (BKA) are worth highlighting. In 1992, they developed a model to assess the potential for organised crime by criminal groups (Meywirth 1999). Assessment models deriving from the methodology designed by the *Crime Research Group* of the University of Ghent (Black et al. 2000, 2001) have also been developed in countries such as Australia or the USA. On the other hand, there are two other mixed models that combine assessing organisations and the markets they develop. One is the Sleipnir project developed by the Royal Canadian Mounted Police (1999) and the other by Klerks (2000). They were both developed using information from a Delphi survey, which revealed some indicators that could be used to measure risk quantitatively.

Finally, there are other models that assess the impact of organised crime and use the assessment of structural or facilitating factors which such crime generates and inspires as a reference. Of all the studies analysed, the Albanese method is particularly remarkable: the Organised Crime Risk Assessment Tool (Albanese 2002) comprises seventeen variables. Also, worth noting is the Risk Assessment Matrix (RAM) (Vanden Beken). This method identifies potential sources of threat, enabling risk comparisons and predictions to be carried out on the matter.

If measuring this phenomenon and its assessment have attracted the attention of the scientific community, the explanations for this typology have also been an essential feature of academic debate. The following section outlines the main explanations for this typology divided into different levels of analysis, which we need to take into account before providing the relevant explanations.

5 Explanations for Organised Crime: Scope of the Different Levels of Analysis

The explanations regarding organised crime have gone through several phases and fields. Depending on the methodological approach and the field of choice, more emphasis has been placed on social, economic or political conditions, which facilitate or explain the emergence of organised crime. The first approximations were descriptive models rather than scientific theories (von Lampe 2011), which developed in a unsystematic manner while attempting to find an explanation for the different aspects of organised crime. However, trying to find a model that comprises the factors influencing the emergence and survival over time of an organised crime group or organisation would be too unfeasible; thus, we will try to divide the different explanatory perspectives into different analytical levels so that we can clearly distinguish, what is being explained and how it is being done. We will use three approaches to group the most relevant explanations concerning organised crime and avoid being overly exhaustive.

5.1 Macro-sociological Level

This explanatory level should reveal political and economic factors that facilitate the development of organised crime in a specific location. A criminal phenomenon does not emerge out of the blue, but rather results from the simultaneous existence, to a greater or lesser extent, of specific factors that enable its development. Even though this level is necessary to explain organised crime, it has received less scientific attention in the literature and it is difficult to find studies that conduct a multi-causal analysis, while finding justification in a number of factors of an economic, political and social nature.

The explanatory models or paradigms as Albanese (2011) likes to call them have centred on explaining criminal organisations mainly from three angles: the hierarchical or bureaucratic model, which explains organised crime based on the structure of organisations as the essence of the explanation and from which an explanatory model centred on the Italian-American mafia can be developed (Cressey 1979); the local and ethnic model centred on culture and ethnicity as a way of explaining organised crime and, finally Smith (1980), and his business model, which explanation centres on the illicit activities and the primacy of the market economy over the group's structure, assimilating the organisation to a legal business (Smith 1980; Reuter 1983).

Very few studies address the factors that favour the emergence and development of organised crime in a specific location and how movements in such development can be detected: organisation displacement, diminishing illicit markets, inter-organisation alliances, etc. In this regard, the work of Albanese (2000) is especially relevant as he proposes a model that uses both opportunity and offender-availability factors to predict the incidence of organised crime activity. Two studies regarding group mobility have been published more recently. Morselli et al. (2010) identified push and pull factors that provide an insight into how and why criminal groups, organisations or general patterns of organised crime are present across a variety of settings. This study highlights the push and pull factors that favour group mobility in various dimensions: relating to criminal markets, criminal groups and conditions in legitimate settings. These type of studies are scarce but absolutely necessary because of their impact on whichever prevention policies and strategies are to be developed. Varese has also published a book on the mobility of mafias in the same direction, explaining the unintentional and intentional factors that account for the presence of the mafiosi in a new territory (Varese 2011).

5.2 Organisational Level

The next explanatory level of organised crime draws on organisation as the subject of the study and is the most explored level if we consider the number of investigations that can be included in this level. The explanatory models or paradigms mentioned above have centred on explaining criminal organisations, focusing mainly on two aspects:

(a) The internal study of criminal organisations as a way of finding out more about their dynamics and internal cohesion mechanisms, which allows them to function underground.

In this regard, the first study that analysed the workings of organisations from within their structure was Cressey (1979), who came up with a hierarchical model based on the testimonies of Valachi, a repentant of the Cosa Nostra who exposed the organisation's best kept secrets. Later, Albini (1971), who was not convinced by this model, came up with an alternative model called *Patron-client model*, which is a more reticular model based on protection and the loyalty relationships of members of the mafia. Still on the subject of structural aspects of organisations, studies have also analysed the size, centralisation and formality of organisations (Sutherland and Potter 1993; Varese 2001; Zaitch 2002; Paoli 2003).

Currently, organisational structure has lost ground on the basis that there are different typologies within a continuum that runs from hierarchical and traditional centralised organisations to reticular structures, which are much more flexible and fluid for protection purposes in more supervised and controlled environments (UNODC 2002). In this way, there is widespread consensus in the literature, through various works that offer empirical support on the subject, that organisations tend to be more flexible and reticular due to improvements in communication, globalisation and the new technologies available, which facilitate a more fluid and not so structured relationship (Kleemans and Van de Bunt 2008; Bruinsma and Bernasco 2004; Natarajan 2006; Morselli and Giguère 2006; Morselli and Roy 2008; Morselli 2009).

These changes have also altered how organisational explanations are understood and addressed and have given rise to a line of investigation that offers investigative and operative advantages over traditional conceptions. Referring to the social network analysis methodology enables to more easily confine the different criminal organisations, from simple and flexible groups of criminals to large structured and permanent criminal corporations (Morselli 2009). This methodological approach originates from the work of the psychologist Moreno (1953) and has been applied to multiple fields, including the study of *covert networks*, which internal working is still unknown due to the caution with which these organisations operate. This tool is a useful and efficient way of knowing more about underlying structures, the most significant players in the group on the basis of their relationship with other members and the most efficient way of dismantling it (Morselli 2009; Bruinsma and Bernasco 2004).

(b) Understanding group influence when explaining individual crime.

Within this type of criminal phenomena, which main feature is criminal activity carried out within the scope of an organisation, the criminal group represents a well-defined criminological factor, which has an vital influence on attracting new members, retaining current members and developing codes and internal rules that justify and facilitate violence and crime. Sutherland (1937) was the first to address the criminological effect of the influence of social interaction with criminals. This impact has been much studied in juvenile delinquency, especially with regard to juvenile gangs. Such is the case of the authors, who studied the effect of criminal group interaction on gaining skills, experience and criminal opportunities (Burt 1992; Granovetter 1973; Lin 2001). However, these developments have not come in the area of organized crime, in which group pressure and influence would be especially relevant, making the group more complex and structured.

On the other hand, through this explanatory aim, some of the traditional criminological explanatory theories have been adjusted to organised crime. This is the case of the differential association theory developed by Sutherland (1947), which ascribes to the criminal group an essential influence on learning criminal behaviour, its techniques and methods and even on the indispensable justification to continue with the criminal behaviour into the future. The differential-opportunity theory developed by Cloward and Ohlin (1960) attributes to the group the crucial opportunity for young people with little prospect of achieving economic and social success in the legal community. This theory, when applied to organised crime, provides an insight into how criminal organisations established in underprivileged neighbourhoods are the main route to prosperity for young people who lack the resources and opportunities to have a future profession in the legal community.

With regard to the magnetic effect that these organisations have for underprivileged young people, Ianni (1974) through the *ethnic succession* theory already recognised that organised crime in the USA was a way of progressing socially (*queer ladder of mobility*) (Bell 1953) for immigrants, who had just set foot on American soil. Currently, these premises still apply and immigration can be a risk factor in relation to involvement in organised crime, not so much because of the nationality element, but because of the lack of legal opportunity conditions for immigrants in the countries of destination. An investigation carried out in Spain on a sample of 1156 members of 72 criminal organisations showed, after analysing their employment status in Spain, that 70% of foreigners did not have a legitimate job. Conversely, this percentage is reversed for Spanish citizens, as only 40% of them were unemployed (Gimenez-Salinas et al. 2009). Generally, the foreign citizens did not have a residence permit and were illegally residing in Spain, and thus, they had limited chances of opting for legitimate employment in comparison with the offers from criminal organisations.

These theories are a transposition of the traditional criminological theories to organised crime, but there are few contributions that explain the opportunities offered by a criminal organisation to adult criminals. The majority are explanations for juvenile delinquency adapted to adult delinquency. Future research into risk factors for this type of delinquency and the factors that facilitate discontinuance should be carried out from an organisational perspective (i.e. how can we work against the criminogenic effect of the organisation and offer legal opportunities in return).

5.3 Individual Explanations

Literature on organised crime is copious in the study of criminal organisations and their internal dynamics, but there is little tradition of empirical studies that address the profile of members of organisations. However, in recent years there have been higher number of investigations on samples of subjects involved in organised crime which address the differences between this type of criminal and common criminals.

The psychological profile of this type of criminal has undergone little research, much less than in the field of white-collar crime, which offers more investigations. Essentially, the only existing investigation was conducted by Bovenkerk (2000) on the profiles of the leaders of organisations, assimilating them to the positions in legitimate companies. This study suggests that the distinctive individual personality traits of leaders of criminal organisations are similar to personality traits that predict leadership success in legitimate business: extraversion, controlled impulsiveness, a sense of adventure or megalomania and Narcissistic personality disorder. There are no other studies in this regard, just other authors who have addressed the abilities, attitudes and skills required of criminals involved in criminal organisations (Beare 1996; Van Duyne 2003; Homer 1974; Levi 1998).

Despite the lack of especially conclusive studies on individual profiles, the latest studies on profiles and criminal careers of organised criminals reveal that the profile of this type of criminal is very different to that of a common criminal and that there are different profiles associated with different functions within a criminal organisation. The profile of a member in charge of group coordination or laundering illegal money is not similar to another member who is in charge of logistical tasks, control or inflicting violence. Even though research is rather scarce, we can provide the conclusions reached with regard to these two points.

(a) Differences with common criminals

Some recent research suggests that there are significant differences between common criminals and organised crime, to the extent that the main theoretical conclusions especially developed by developmental criminology, would not apply to organised criminals. The main differences relate to gender and age. Female delinquency is more present in organised crime (17% compared to 10.8% in common delinquency) than common crime (Gimenez-Salinas et al. 2011). Regarding criminal age, in average is higher in organised crime (an average of between 30 and 40 compared to 18–21 in common delinquency) (Gimenez-Salinas et al. 2011; Van Kopper et al. 2010a). These conclusions are in line with the skill set required to be a member of an organised crime organisation. The organisational functions of this type of crime entail such planning, experience, specialisation and high degree of comingling with legitimate companies that they require skills, abilities, know-how and contacts which are typical of a much more adult profile, with professional experience, and one that is very different to that of a common criminal (Gimenez-Salinas et al. 2011).

The involvement of foreigners in criminal groups is also a trait of this type of criminality, which has a bearing on the individual profile. For this reason, the composition of groups has traditionally been ethnically homogenous (Italian, Chinese, Russian mafias, etc.). Many reasons explain the incorporation of members with similar nationality: loyalty and trust (Enderwick 2009); a way of extending its tentacles to other territories located abroad; an easy way for recruitment: illegal residents are much more in need of illegal resources to survive in the country (De La Corte and Gimenez-Salinas 2010).

Looking further into the differences with common criminals, inconsistencies with the main assumptions of development criminology with regard to common crimes are found. The main theories (Sampson and Laub 2005a, b; Moffit 1993, etc.) claim that the vast majority of common criminals are most productive at a young age, and their criminal career decreases as they get older and begin to assume conventional adult commitments, such as marriage or a legitimate job, which will act as facilitators for desistance. With regard to organised criminals, this pattern does not apply, since they are adult criminals, who lead normal lives for people their age—most are married or are in a stable relationship with children—and, generally, have a legitimate job (Requena et al. 2014). On the whole, the factors facilitating desistance that apply to common criminals do not apply to organised crime members.

On the other hand, another basic position which results from studying the criminal career of common criminals is that the bulk of them (80–90%) have a short career concentrated in the period of delinquency; the rest would be very highly productive subjects in terms of delinquency and a lasting career. Van Koppen et al. (2010a, b) studied 854 subjects from organised crime groups and concluded that their criminal career did not follow this pattern, but rather that a large number of subjects started their criminal career as adults and did not have any prior police records. According to this study, the criminal careers of members of criminal organisations were found to follow four criminal trajectories: a group of adult-onset offenders (40%) and a group without any previous criminal records (19%), a group of early starters (11%) and a group of persisters (30%).

(b) Typologies Depending on Organisational Functions

As mentioned above, a major finding from recent research is that one cannot just speak about a single profile of organised crime and that diversity can be explained by the different functions carried out within organisations. In the same vein, Kleemans and De Poot (2008) identified two typologies of organised criminals depending on the functions and skill set required to operate in the organisation, which we can extend to three according to some recent studies:

(a) The first typology includes the *adult criminal* profile—aged between 28 and 49—who have a structured social and personal life—with a legitimate job and a stable family life with children—who starts his or her criminal career late, and, normally, has no criminal records. The subjects often come from the legitimate economy, in which their profession offers a window of opportunity into the world of crime, due to their experience and expert knowledge or because they have valuable contacts for the illicit trade (Kleemans and De Poot 2008; Piquero and Benson 2004; Requena 2011). This type of subject tends to occupy more complex functions in the organisation: those requiring coordination or specific know-how. The integration of this group into organised crime seems to be caused by the pursuit of financial benefit in addition to their income from their legitimate job (Finckenauer 2005).

(b) The second typology includes the *common criminal* profile, which Kleemans and De Poot (2008) refer to as the *local hero*. This subject is a criminal who becomes involved in common criminal activities at an early age and he or she is able to move into organised crime because of his or her experience in the field. Motivation which drives these individuals' behaviour in organised groups is the same as for common crimes, that is to achieve the social status and financial benefits that are beyond their reach in the legal market (Cloward and Ohlin 1960; Requena 2011). These types of criminals carry out the most basic functions in the organisation: operation, transport, logistics, etc., and organised crime is a way for them to reach the pinnacle of their criminal career with greater economic returns. This type of profile also encompasses the highest concentration of foreign citizens who reside in the country illegally, and whose involvement in crime is probably the only way for them to make money in a country where they have no access to legitimate employment (Gimenez-Salinas et al. 2011; Ianni 1974).

(c) A third typology is that formed by the family members of those involved in crime groups, which on a more or less permanent basis, form part of the group. In these cases, the transition to organised crime is enhanced by their family link, which ties provide more or less stable opportunities of collaboration (Gimenez-Salinas and Fernandez Regadera 2016). In this case, the presence of women is overrepresented, since they offer essential logistical and material support as facilitators of the criminal business (Morselli 2009), but remain in the background because this is secondary to the organisation's main functions.

6 Conclusions

As shown in this chapter, the concept of organised crime is complex and comprises numerous dimensions, all of which are relevant for its existence and development. Scientific research in the field has gone through numerous phases aimed at knowing more about a phenomenon that is difficult to encompass in full and regarding which there was little consensus in the 1980s and 90s. Currently, we can say that we have now overcome that phase of uncertainty and we now know the problem we are facing well. However, we are still lacking numerous key insights to be able to design efficient preventative and repressive mechanisms. With regard to measuring, we lack the instruments to provide a reliable international measurement to correctly measure this phenomenon's multiple dimensions periodically. The existing measuring tools are limited and address specific dimensions and not organised crime in full.

From an explanatory perspective, there are still many unanswered questions, we need to analyse macro-social theories in depth to identify the factors that facilitate organised crime so that they can ultimately become effective measures for prevention and repression. In future, we should further explore the factors that facilitate this type of crime and the factors that facilitate discontinuance from an organisational point of view. Only in this way can we thwart the criminogenic effect of the organisations and offer legal opportunities of change.

From an individual point of view, we need to make more progress on the explanations applicable to the criminal typology characteristic of adults, which is used as financial cover by many families who are dedicated to crime "professionally". These explanations should identify the appeal, continuance and discontinuance factors of the criminal career of these subjects. We now know that it is different to common crime and that traditional inferences regarding criminology far from effectively explain this crime typology. This is the right time to widen the search in this field and provide adequate responses to what are the risk and protection factors for this type of crime, or what are the criminal careers for organised crime members like, or which are the most effective desistance factors for that type of professional criminality. Only by successfully answering these types of questions, it can be considered to reduce this activity and devise efficient alternatives to this type of crime.

References

Albanese, J. (2000). The causes of organized crime: Do criminals organized around opportunities for crime or do criminal opportunities create new offenders? *Journal of Research in Crime and Delinquency, 38*, 316–319.
Albanese, J. (2002). *The prediction and control of organized crime: A risk assessment instrument for targeting law enforcement efforts*. Richmond, VA: Virginia Commonwealth University.
Albanese, J. (2008). *Criminal justice* (4th ed.). Boston: Allyn & Bacon.
Albanese, J. (2011). *Organized crime in our times* (6th ed.). New York: Elsevier.
Albini, J. (1971). *The American Mafia: Genesis of a legend*. New York: Appleton-Century-Crofts.
Arlacchi, P. (1986). *Mafia Business. The Mafia ethic and the spirit of capitalism*. London: Verso.
Beare, M. (1996). *Criminal conspiracies: Organized crime in Canada*. Scarborough, Ontario: Nelson.
BEEPS. (1996). *Business environment and enterprise performance survey*. World Bank ad the EBRD.
Bell, D. (1953). Crime as an American way of life. *The Antioch Review, 13*(2), 131–154.
Black, C., Vander Beken, T., & De Ruyver, B. (2000). *Measuring organised crime in Belgium: A risk-based methodology*. Antwerp: Maklu.
Black, C., Vander Beken, T., Frans, B., & Paternotte, M. (2001). *Reporting on organised crime. A shift from description to explanation in the Belgian Annual Report on organised crime*. Antwerp: Maklu.
Block, A. A., & Chambliss, W. J. (1981). *Organizing crime*. New York: Elsevier.

Bovenkerk, F. (2000). Wanted mafia boss: Essay on the personology of organized crime. *Crime, Law and Social Change, 33*(3), 225–242.
Bruinsma, G., & Bernasco, W. (2004). Criminal groups and transnational illegal markets: A more detailed examination on the basis of social network theory. *Crime, Law and Social Change, 41* (1), 79–94.
Burt, R. S. (1992). *Stuctural holes: The social structure of competition*. Cambrige, MA: Harvard University Press.
Catanzaro, R. (1991). *Il delitto come impresa. Storia sociale della mafia*. Milan: Rizzoli.
Cloward, R. A., & Ohlin, L. E. (1960). *Delinquency and opportunity: A theory of Delinquent Gangs*. New York: Free Press.
Cressey, D. R. (1979). *The theft of the Nation: The structure and operations of organized crime in America*. New York: Harper and Row.
De La Corte, L., & Gimenez-Salinas, A. (2010). *Crimen.org. Evolución y claves de la delincuencia organizada*. Barcelona: Ariel.
Dugato, M. (2013). *The crime against businesses in Europe: A pilot survey*. Europe: Transcrime and Gallup Organisation.
Edward, A., & Gill, P. (1999). The politics of "transnational organised crime": Discourse, reflexivity and the narration of threat. A versión of this paper presented to the UK Economic and Social Research Council Research Seminar Series on Policy Responses to Transnational Organized Crime. University of Leicester, September 8.
Enderwick, P. (2009). Applying the eclectic framework: The strategy of transnational criminal enterprises in the global era. *Critical perspectives on international business, 5*(3), 170–186.
Finckenauer, J. O. (2005). Problems of definition: What is organized crime? *Trends in Organized Crime, 8*(3), 63–83.
Finckenauer, J. O. (2010). *Mafia y crimen organizado. Todo lo que hay que saber sobre la Mafia y las principales redes criminales*. Madrid: Península.
Gimenez-Salinas, A, De La Corte, L., Requena, L., & De Juan, M. (2009). La medición y evaluación de la criminalidad organizada en España: ¿Misión Imposible? *Revista Española de Investigación criminológica, 7*.
Gimenez-Salinas, A., & Fernandez Regadera, S. (2016). Multiple affiliation in criminal organizations: Analysis of a Spanish simple. *Crime, Law and Social Change, 65*(1), 47–65.
Gimenez-Salinas, A., Requena, L., & De La Corte, L. (2011). ¿Existe un perfil de delincuente organizado? Exploración a partir de una muestra española, *Revista Electrónica de Ciencia Penal y Criminología*, 13-03.
Granovetter, M. (1973). The strength of weak ties. *American Journal of Sociology, 78*(6), 1360–1380.
Hagan, F. E. (1983). The organized crime continuum: A further specification of a new conceptual model. *Criminal Justice Review, 8*(2), 52–57.
Hagan, F. E. (2006). Organized crime and organized crime: Indeterminate problems of definition. *Trends in organized crime, 9*(4), 127–137.
Hawkins, G. (1969). *God and the mafia*. New York: National Affairs Inc.
Hobbs, D. (2001). The Firm. Organizational logic and criminal culture on a shifting terrain. *The British Journal for Criminology, 41*(4), 549–560.
Homer, F. D. (1974). *Guns and garlic: Myths and realities of organized crime*. West Lafayette, Indiana: Pordue Univesity Press.
Ianni, F. A. J. (1974). *Black Mafia: Ethnic succession in organized crime*. New York: Simon & Schuster.
Kaufmann, D., Kraay, A., & Mastruzzi, M. (2003). *Governance matters III: Governance indicators for 1996-2002*. Policy research working paper 3106. World Bank: Washington, DC.
Kleemans, E. R., & De Poot, C. J. (2008). Criminal careers in organized crime and social opportunity structure. *European Journal of Criminology, 5*(1), 69–98.
Kleemans, E. K., & Van de Bunt, H. G. (2008). Organised crime, occupations and opportunity. *Global Crime, 9*(3), 185–197.

Klerks, P. (2000). *Groot in de hasj: theorie en praktijk van de georganiseerde criminaliteit*. Doctoral thesis. Erasmus Universiteit Rotterdam.
Landesco, J. (1929). Organized crime in Chicago. *The illinois crime survey* (pp. 823–1087). Chicago: The Illinois Association for Criminal Justice.
Lashly, A. V. (1930). The illinois crime survey. *Journal of the American Institute of Criminal Law and Criminology, 20*(4), 588–605.
Levi, M. (1998). Organising plastic fraud: Enterprise criminals and the side-stepping of fraud prevention. *The Howard Journal, 37*(4), 423–438.
Lin, N. (2001). *Social capital: A theory of social structure and action*. Cambridge: Cambridge University Press.
Merchant International Group Ltd. (MIG). (2003). *Grey area dynamics, organized crime figures 2004: Special analysis for UNICRI Turin*.
Meywirth, C. (1999). Das Lagebild Organisierte Kriminalität. *Kriminalistic, 53*, 447–452.
Moffit, T. E. (1993). Adolescence-limited and life-course persistent anti-social behavior: A developmental taxonomy. *Psychological Review, 100*, 674–701.
Moore, W. H. (1974). *The Kefauver Committee and the politics of crime, 1950-1952*. Columbia: University of Missouri Press.
Moreno, J. L. (1953). *Who shall survive? Foundations of sociometry, groups psychotherapy and sociodrama*. New York: Beacon House Ltd.
Morselli, C. (2009). *Inside criminal networks*. New York: Springer.
Morselli, C., & Giguère, C. (2006). Legitimate strengths in criminal networks. *Crime, Law and Social Change, 45*(3), 185–200.
Morselli, C., & Roy, J. (2008). Brokerage qualifications in ringing operations. *Criminology, 46*(1), 71–98.
Morselli, C., Turcotte, M., & Tenti, V. (2010). The mobility of criminal groups. *Global Crime, 12*(3), 165–188.
Natarajan, M. (2006). Understanding the Structure of a large heroin distribution network: A quantitative analysis of qualitative data. *Journal of Quantitative Criminology, 22*(2), 171–192.
Paoli, L. (2002). The paradoxes of organized crime. *Crime, Law & Social Change, 37*(1), 51–97.
Paoli, L. (2003). The invisible hand of the market: The illegal drugs trade in Germany, Italy and Russia. In P. van Duyne, K. von Lampe, & J. Newell (Eds.), *Criminal finances and organising crime in Europe* (pp. 19–40). Nijmegen, Netherlands: Wolf Legal Publishers.
Piquero, N. L., & Benson, M. L. (2004). White collar crime and criminal careers. Specifying a trajectory of punctuated situational offending. *Journal of Contemporary Criminal Justice, 20*(2), 148–165.
Requena, L. (2011). *Delincuencia organizada: perfil criminológico de una muestra de miembros activos en organizaciones criminales que han actuado en España entre 1999 y 2010*. Tesis doctoral no publicada: Universidad Autónoma de Madrid.
Requena, L., De Juan, M., Gimenez-Salinas, A., & De la Corte, L. (2014). A psychosocial study on crime and gender: Position, role and status of women in a sample of Spanish criminal organizations. *International Journal of Social Psychology, 29*(1), 121–149.
Reuter, P. (1983). *Disorganized crime: The economics of the visible hand*. Cambridge, Mass: MIT Press.
Reuter, P. H., MacCoun, R. J., Murphy, P., Abrahamse, A., & Simon, B. (1990). *Money for crime: A study for the economics of drug dealing in Washington, D.C.* Santa Monica, C.A.: Rand Publication Series.
Royal Canadian Mounted Police. (1999). *Sleipnir: The long mattress for organised crime, an analytical technique for determining relative levels of threat posed by organised crime groups*. Criminal Analysis Branch, Criminal Intelligence Directorate.
Ruggiero, V. (1996). *Organized and corporate crime in Europe: Offers that can't be refused*. Aldershot: Dartmouth.

Ruggiero, V. (2000). Transnational crime: Official and alternative fears. *International Journal of the Sociology of Law, 28,* 187–199.

Sampson, R. J., & Laub, J. H. (2005a). Life-course desisters? Trajectories of crime among delinquent boys followed to age 70. *Criminology, 41*(3), 301–339.

Sampson, R. J., & Laub, J. H. (2005b). A life-course view of the development of crime. *Annals of the American Academy of Political and Social Science, 602,* 12–45.

Senate, U.S. (1951). *Special committee to investigate organized crime in interstate commerce.* Final Report of the Special Committee to Investigate Organized Crime in Interstate Commerce (Vol. 82).

Smith, D. C. (1976). Mafia: The prototypical alien conspiracy. *The Annals of the American Academy, 423*(1), 75–88.

Smith, D. C. (1980). Paragons, Pariahs, and pirates: A spectrum-based theory of enterprise. *Crime & Delinquency, 26*(3), 358–386.

Sutherland, D. (1947). *Principles of criminology* (4th ed.). Philadelphia: Lippincott.

Sutherland, E. H. (1937). *The professional thief: By a professional thief.* Chicago: The University of Chicago Press.

Sutherland, M. D., & Potter, G. W. (1993). Applying organization theory to organized crime. *Journal of Contemporary Criminal Justice, 9*(3), 251–267.

Trasher, F. M. (1927). *The gang: A study of 1313 in Chicago.* Chicago: University Press.

UNODC. (2002). *Results of pilot survey of forty selected organised criminal groups in sixteen countries.* Available at: http://www.unodc.org/

UNODC. (2007). Alvazzi del Frate, A. *Study on Crime and Corruption in Cape Verde.* Available at the website: https://www.unodc.org/documents/data-and-analysis/dfa/Study-crime-corruption-english.pdf

van Dijk, J. (2007). Mafia markers assessing organized crime and its impact upon societies. *Trends in Organized Crime, 10*(4), 39–56.

Van Duyne, P. C. (2003). Organizing cigarrette smuggling and policy making ending up in smoke. *Crime, law and Social Change, 39*(3), 285–317.

Van Koppen, M. V., De Poot, C. J., & Blokland, A. J. (2010a). Comparing criminal careers of organized crime offenders and general offenders. *European Journal of Criminology, 7*(5), 356–374.

Van Koppen, M. V., de Poot, C. J., Kleemans, E. R., & Nieuwbeerta, P. (2010b). Trajectories in organized crime. *British Journal of Criminology, 50*(1), 102–123.

Vander Beken, T. (2004). Risky business: A risk-based methodology to measure organized crime. *Crime, Law and Social Change, 41*(5), 471–516.

Varese, F. (2001). *The Russian Mafia: Private protection in a new market economy.* Oxford: Oxford University Press.

Varese, F. (2011). *Mafias on the move: How organized crime conquers new territories.* Pinceton, NJ: Princeton University Press.

Vazquez González, C., & Serrano, M. D. (2007). *Derecho Penal Juvenil.* Madrid: Dykinson.

von Lampe, K. V. (2001). Not a process of enlightenment: The conceptual history of organized crime in Germany and the United States of America. *Forum on Crime and Society, 1*(2), 99–118.

von Lampe, K. V. (2002). Organized crime research in perspective. In P. C. Van Duyne, K. von Lampe, & N. Passas (Eds.), *Upperworld and underworld in cross-border crime.* The Netherlands: Wolf Legal Publishers.

von Lampe, K. V. (2004). Measuring organized crime, a critique of current approaches. In P. C. Duyne et al. (Eds.), *Threats and phantoms of organised crime, corruption and terrorism; rhetoric and critical perspectives.* Nijmegen: Wolf Legal Publishers.

von Lampe, K. V. (2011). The use of models in the study of organised crime. Paper presented at the 2003 conference of the European Consortium for Political Research. Marbourg: Germany.

von Lampe, K. (2016). *Organized crime. Analyzing illegal activities, criminal structures and extra-legal governance*. Thousand Oaks: Sage.
World Economic Forum. (2003). *The global competitiveness report 2003-2004*. New York: Oxford University Press.
Zaitch, D. (2002). *Trafficking cocaine: Colombian drug entrepreneurs in the Netherlands*. The Hague: Kluwer Law International.

Measuring Organised Crime: Complexities of the Quantitative and Factorial Analysis

Daniel Sansó-Rubert Pascual

1 Relevance and Impact of Discourses on Organised Crime for Measurement

Since the late twentieth century and early twenty-first century, organised crime has been the object of a progressive inclusion into the list of threats and risks to security (Sansó-Rubert Pascual 2008) and even to defence (Ruiz-Funes García 1960), by virtue of the increasing links between it and international conflicts (Ibáñez Muñoz and Sánchez Avilés 2015), as well as because of the repercussions that the development of its multiple criminal activities has on the integrity and independence of States in economic, political and social terms (Sansó-Rubert Pascual 2011).

On the one hand, the recognition of the danger it poses has permeated a good deal of political, institutional and academic discourses (García Segura 2015: 207) for different purposes. On the other hand, it is also possible to identify discourses that are quite openly against labelling as a threat the very nature of organised crime (Edwards and Gill 2002: 246), and even detractors who question the very existence of organised crime, at least in the terms institutionalised through the United Nations Convention against Transnational Organised Crime and its protocols, signed in Palermo in 2000 (Fernández Steinko 2008).

Each of these discourses is not harmless, putting forward different approaches and perspectives based on specific interests. One can deduce that those that reject

---

D.S.-R. Pascual (✉)
Security Studies Center (CESEG), University of Santiago de Compostela, Santiago de Compostela, Spain
e-mail: daniel.sanso-rubert@usc.es

the interpretation of organised crime in the terms that are mainly accepted nowadays do so due to the imposition of certain countries and institutions. A political construct has colonised the collective imagination of the public, alongside the national and international laws and policies of governments and international organisations (Fernández Steinko 2008: 61–87).

It is a cultural inertia strongly inspired by the institutional interests of some nations, like the USA, eager to channel the efforts in the field of security and defence to new threats, that justifies the investment made and its future sustainability, promoting the idea of organised crime as a potential danger that hangs over the whole of society. This chapter also argues that behind the strengthening of the global fight against organised crime lies a political discourse that pretends to hide the pursuit of hegemony, the control of the international security and defence agendas and the defence of explicit economic interests, linked to the sectors that are enriched with the fight against crime (Neuman 1997). The background is the control and imposition of economic and financial rules but also those related to security. Susan Strange has called this form of exercise of power, "structural power", an indirect form of exercising hegemony over what strategists call "soft power" (Nyer 2004). In short, this is what Díez Ripollés (2003) calls "legal colonisation", a process whereby a country has the capacity to impose to another country's legal system—or to the international community as a whole—its particular way of defining and diagnosing a phenomenon, in order for it to become laws and legally binding documents. It is a process that can be extended to the unilateral diagnosis of threats to national security of countries around the world. A sort of "legal acculturation" is to borrow a term from cultural anthropology (Ronderos 2005: 208).

From a political perspective, an attempt is made to justify the harmful genesis of organised crime—with or even without empirical verification (Reuter 2001)—to articulate response protocols and resource mobilisation in order to justify extraordinary measures which may involve, if necessary, the collapse of the balance between security and civil liberties and fundamental rights. This dubious empirical approach (evidence-based approach) of the analysis and crime control policies is a matter of much academic discussion. The criticism is that public policies against organised crime are not based on the result of empirical research but are the product of the system of normative beliefs and coalitions of interests that determine them: the interested politicisation of the discourse against organised crime (discourses like "war against organised crime", "zero tolerance" or "iron fist policies"), for the sake of achieving political gains.

There is ample evidence of the use of questionable data, it being the object of speculation and intentional distortion (Andreas 2011: 23–25), to justify certain policies (Walsh 2004: 9), or to gain the attention of the international community about certain aspects of organised crime, like illicit flows.

Organised crime has become a matter of high political sensitivity (high politics), leading to the alteration and misrepresentation of data in support of political and ideological positions (Thoumi 2005: 187), linking the success or failure of States (in reality, governments), directly with their capacity to provide security.

The existing discursive problem, therefore, affects research on organised crime. Data collection and measurement are seriously affected by the interests of the actors who use them because before one can approach the merely technical matter of how to measure something, one has to previously know what to measure, depending on the nature of the problem and the objectives at stake (political, institutional, economic, moral or ideological interests), thereby helping or hindering, depending on the case, academic and empirical research.

It is therefore important to place the discourse of organised crime and how to deal with it in the context of criminological research (academic discourse), oriented towards the implementation of public policies; prioritise the investigation and scientific knowledge on the subject above any type of interest that directly influences the approaches, strategies and methodological analysis applied to the study of the matter and that could ultimately distort the measurement of organised crime.

In conclusion, the academic discourse should prevail—even though it is not exempt from deficiencies—in order to use contrasted empirical knowledge on the reality of organised crime, in an attempt not only to gain knowledge about the present reality on the subject of organised crime but also to establish future trends and to be able to explain the success or failures of these organisations and criminal activities, including aspects such as mobility, structures, composition, capabilities, types of relationships or possible links with other phenomenologies such as terrorism or insurgencies (Pulido Gragera and Sansó-Rubert Pascual 2014).

2 The Use of Criminology as an Academic Discipline Suitable for the Study and Measurement of Organised Crime: A Brief Critical Reflection

The last two decades have been characterised by an intense doctrinal debate on the definition of organised crime, finally settled unsatisfactorily by international law in favour of the recognition of an own entity and phenomenology (Abadinsky 2004; Albanese 2001; Sansó-Rubert Pascual 2008).

The international scene forms a ragtag group of organisations whose structure, discipline, internal rules, division of roles, illegal activities and thus their dangerousness represent a plurality of combinations (De la Corte and Giménez-Salinas Framis 2010). This diversity is precisely the main impediment to outline a universal definition that manages to capture the essence and common variables of the totality of these criminal manifestations.

Despite the inherent difficulty that a multifaceted threat poses because of its diffuse and elusive nature, we pretend to outline a scheme of the progress made in the measuring of transnational organised crime, emphasising its most significant elements. The objective is to help create a comprehensive analytical framework within which both the conceptual dimension and aetiology are discussed. Not forgetting that organised crime, like any other social phenomenon, is closely related to the realities that surround it. Crime does not take place in the "abstract" but takes place in particular circumstances of time and space. Social conditions, technological, political and human development, have a decisive influence in the way that crime takes place, in the forms it manifests itself, in its quantity, intensity and in all its connotations and peculiarities (Sansó-Rubert Pascual 2008). Knowledge of these aetiological factors also provides significant inputs to criminological studies. Therefore, the need to understand what is organised crime, the possible paths through which it evolves and its main manifestations and effects raises a number of questions that, logically, require answers. Answers that can fill the shortcomings of current scientific knowledge, especially those organised criminal manifestations with a capacity to wrestle territorial and social control from States, but also the monopoly of violence (Sansó-Rubert Pascual and Giménez-Salinas Framis 2014); for this purpose, criminology is an indispensable tool. It has a great, untapped, explanatory potential that requires higher levels of investment in criminological investigation projects. This scientific discipline should be the recipient of a higher confidence on the part of the actors that could benefit from it: an idea that Bernard et al. (2010: 337–38) have reflected very well with these words:

> In the past, criminal policies have frequently been the product of political ideology: the conservatives favoured certain policies and the progressives favoured others. Neither were particularly interested in investigating whether the policies adopted worked. Instead, quite frequently, they seemed to take the following position: "I have taken a decision, don't bother me with the facts" (…). In the future, criminological theory and investigation, more than political ideology, should be the main source of criminal policy.

It seems logical that the policies adopted in the fight against organised crime should take into account "what works". If we disregard this, we run a high risk of leaving society unprotected, unnecessarily sacrificing individual liberties and absurdly squandering public resources (Sansó-Rubert Pascual 2016).

But to reach that point in the path, we still have a long way to go when it comes to training and research. To the effect of its practical application, it also requires the development of a set of analytical methodologies, reinforced with the relevant educational support adapted to the peculiarities of the criminological discipline.

However, despite the aforementioned difficulties within the curriculum of criminology, organised crime is gaining weight as a specific field of study and teaching, reflecting the concerns raised by the criminal phenomenology. That is the reason why the list of issues to consider—and to try to provide a scientific explanation—has only multiplied, demanding, in turn, the development of the appropriate instruments and methodologies for its analysis and explanation.

These circumstances should act as a spur to gain momentum and promote a commitment to the criminological scientific field. The methodological and didactic challenges are present, and as criminologists ascribed to academia, we must accept them and provide answers (Sansó-Rubert Pascual 2016).

3 The Need for a Quantitative Approach to Organised Crime

Is it possible to establish measures for complex criminal demonstrations as it is done with conventional crimes? Is it reasonable to use the same measuring instruments? What obstacles can we find when we measure their size and get to know their dimensions? These are rotating questions in the criminological doctrine that is recently fostering studies on organised crime beyond entirely qualitative considerations (Van Dijk and Buscaglia 2003; Giménez-Salinas Framis et al. 2009; Sansó-Rubert Pascual 2012). There is an increasing academic interest to boost the actual deficient scientific method on the study of organised crime development through the use of quantitative analytical methodologies.

Although certainly the amount of studies, books, articles in scientific magazines and reports on organised crime have experienced a constant increase during recent years, the balance of knowledge reveals a highlighted disproportion in favour of qualitative researches (descriptive), to the detriment of the ones entirely quantitative. This is why the interest and efforts of some criminologists have been focused on the development of new ways of measurement and assessment (Von Lampe 2004; Vander Beker 2004; Albanese 2000; Giménez-Salinas Framis et al. 2009; Sansó-Rubert Pascual 2012).

The selected case of study and the methodology used for it suffer from an important burden: the opacity inherent to the investigations about organised crime, as well as to the multidimensionality of the concept and the identity of the group which characterises it. The tendency of a criminal subculture towards illegality is a factor that has historically restricted the access to direct sources, that is to say, to the members of organisations involved in illegal activities.

One of the main challenges when attempting to understand organised crime is the debate on measurement and assessment methods. Issues such as the search for more accurate forms of measurement and more rigorous assessment methods, the establishment of more accurate indicators and indexes and the identification of new ways to foster information transfer related to said criminality should be addressed in order to, on the one hand, be able to compare empirically the seriousness and impact of organised crime, and, on the other hand, to be able to develop strategies and policies that suit the specific features and proportions of each individual case (Giménez-Salinas Framis et al. 2009).

Paving the way for quantitative study of organised crime implies opening a sort of Pandora's box, which brings about analytical difficulties regarding the access to sources and methodological gaps. Measuring a concept is intrinsically linked to the composition and nature of the problem itself. The more complex the problem is, the more complex it will be to measure. The complications inherent to any empirical investigations on organised crime have been repeated ad nauseum.

The analysis developed in these pages is an exploratory exercise that aims to encourage further research and debate. Thus, this paper does not intend to be conclusive, but rather to give way to a research and discussion agenda aimed at advancing scientific knowledge, which will enable the tackling of current knowledge gaps around organised crime.

In this regard, there are many study areas within the scope of organised crime that go beyond criminal groups, criminal activities and markets, which contain massive amounts of information that still have to be exploited conveniently. It would be of particular interest to develop analysis on individuals belonging to organisations and their criminal records, as well as developing a new methodology to uncover new aspects of said organisations and how relationships are built on the inside (Requena Espada 2014). Along the same research line, it would also be of great interest to study the group of figures and professions that are identified as facilitators, as well as the typology of relationships inside the organisation and ties between criminal organisations (Morselli 2009). Other aspects subject to analysis might be the knowledge of logistics that criminal organisations need to carry out their activities or the impact of the criminal subcultures on the youth socialisation process and its incidence regarding commonly accepted societal values, as well as minors self-introduction (co-optation), or any other factors seen as contributories to the emergence and expansion of organised crime.

Other relevant areas for criminological research include: determining the presence and distribution of criminal organisations in a certain territory (criminal density); detecting the ability of organised crime to infiltrate the institutional framework as well as its typology (multilevel intensity); investigating new gender approaches (the role of women in organised crime) (Sansó-Rubert Pascual 2010); analysing the replication of criminal behaviour (how and why); analysing those geographical areas and political scenarios that are prone to the emergence of crime (criminal geopolitics); developing early warning and organised crime danger assessment mechanisms (Sansó-Rubert Pascual 2012); determining typologies of identified criminal organisations and their organisational patterns and action methodologies (modus operandi); examining criminal patterns in the business model; and studying the cooperative relationships that are built under the scope of organised crimes (Williams 2005), just to mention a few interesting areas regarding the development of security strategies.

There is a still much to investigate regarding victims of organised crime, despite the increasing academic research on the matter (Spalek 2006; Walklate 2007; Giménez-Salinas Framis 2013). Criminological research has clearly stated that organised crime has such an impact on the social, political and economic sphere that it goes beyond individual victims. Consequently, the hypothesis about the lack of victims in organised crime is completely refuted (Giménez-Salinas Framis 2013). Meanwhile, the main challenge is to determine what criteria should be used to evaluate both direct and indirect damages caused by organised crime, how victims can be identified and how their needs can be identified and guaranteed. As a result of all this, the articulation of prevention strategies that prevent and, if so, reduce damage from arising, should be considered a priority (Sparrow 2008).

One of the biggest challenges posed by organised crime is how to obtain reliable numeric data. Starting from the premise that not every crime is known and that not every known crime is reported, one frequently finds an overwhelming number of scattered estimates that are difficult to contextualise using known methodologies. These figures are reproduced in several publications, with no specific criteria. It is very hard to measure the scope and the volume of organised crime (direct and indirect costs and the amount of revenue generated by legal and illegal criminal activities, as well as the analysis of operational maintenance of criminal organisations), taking into account that the results reflect, in most cases, mere estimates. These indicative figures are useful, but can only be used as estimates.

In line with the previous paragraph, the lack of transparency within the scientific community and its public institutions should be underlined, in regard to the release of assessments in organised crime. As resources and methodology used on papers are not identified, it is not possible to verify their authenticity (Requena 2014). The reliability and value a scientific method brings to any criminological analysis lie on the methodology and empirical research used to enable their verification and comparison. We need to cope with the indispensable need to work with reliable quantitative data in organised crime research. Criminology has achieved this challenge assuming the necessity to commit to methodology complementarities, considering recognised limitations from different criminological existing methods in terms of legitimacy and reliability as the best approach to diverse aspects of organised crime phenomenon as well as a subject of analysis (Fernández Villazala 2008).

Furthermore, we are facing a pressing need to move forward in the development of comparative analyses, now a pending issue in the criminology research agenda. The development of Comparative Criminology will favour the use of a common terminology, as well as allow the determination of the extent to which national structures and cultures influence the scope, types, distribution and characteristics of violence (Howard et al. 2000; Albanese 2008; Requena Espada et al. 2012). This will also analyse the efforts made at tackling criminality, within and between countries (Nelken 1994), as well as sponsor the theoretical development of this discipline beyond national borders (Mueller and Adler 1996). It will give ground to the assessment of national, regional and international security and prevention policies, as well as to the comparison between different countries. It will help to identify the advantages and disadvantages of the many crime fighting systems, so as to come up with relevant strategies (Moore and Fields 1996). To sum up, as exposed decades ago by Bennet and Lynch (1990), transnational crime studies play a crucial role in the drafting of terminology and common political initiatives, a matter of particular importance in the current globalised world and in light of an increasingly transnational crime.

4 Quantitative Data on Organised Crime: Where to Browse and How to Manage Methodological Obstacles for Its Use

Organised crime research pieces usually begin with an introduction, in many cases self-justifying about the results, exposing the difficulties of finding reliable data sources (Resa 1993). A multidimensional concept such as this one creates measuring difficulties, given that the measuring formulae that are normally used for conventional crime are not applicable (Van Dijk 2009).

Another of the challenges faced by criminological research is the establishment of feasible ways of obtaining information sources about organised crime.

Under the conceptual umbrella of "organised crime", it is possible to analyse groups and organisations, activities and/or legal or illegal markets in which they take part (primary activities) and all of the activities that are necessary for the survival of such organisations, considered as inseparable from the main activity (secondary activities).

An initial overview of the resources (data collection) reveals a big range of obstacles that must be correctly handled by the researcher as part of an aim to succeed in the construction of useful indicators for any measurement exercise.

Infourth place, one encounters methodological and information management problems, such as the "dark figure of crime". It is important to keep in mind that the regular sources about crime (such as allegations or police reports) only refer to criminal activities that have been detected by the institutions; thus, they do not reflect criminal activity in its entirety, but only the part that is known to us—as it was previously stated, criminal organisations make great efforts to hide their activities in order for them to go unnoticed.

Another methodological difficulty lies in the classification of criminal activities within criminal law (Alvazzi 2008). Criminal activity, from a criminal law perspective, relates to a specific type of crime, referred to as complex, which are measured by means which are different from those used in conventional offences (Van Dijk 2009). A first approach shows that organised crime does not belong in conventional crime typology, but in a specific category whose measurement requires differentiated indicators. In fact, indicators such as the rate of violent homicides and many others based on conducts which are classified as crimes in Penal Codes do not show whether those manifestations are grounded on organised crime or, conversely, they are mere manifestations of the deteriorating security in some scenarios—such as the Northern Triangle of Central America (Honduras, Guatemala and El Salvador). In those states, access to firearms and their daily use by petty crime perpetrators does not show which homicides are related to organised crime and which are not.

In addition, the codification of criminal activities within criminal law implies a limitation regarding regional and international comparisons. Criminal offences included in each state's national law provide different conceptual definitions for criminal acts, as well as for the circumstances that must accompany the committing attitude for such acts. This circumstance along with the fact that enforceability and validity of each national law is restricted to the national territory of a particular state adds another methodological limitation to take into account.

Von Hofer (2000) describes transnational comparisons of crime as risky on the basis of national crime statistics, given the fact that every country implements different "construction rules". Specifically, the author has identified three main types of factors that have an influence on statistics: substantive factors, legal factors and statistical factors.

Substantive factors depend on the chances that citizens report crimes to the police, on how efficiently the justice system registers those crimes and on the real criminality rate in each country.

Legal factors are related to the different ways of describing a crime and to the characteristics of the subsequent legal proceedings.

Statistical factors refer to the different methods for the production of statistics, i.e. the statistical calculation rules used to gather data about crime. These rules raise great difficulties considering that they determine (a) the way in which data are registered; (b) the computation unit used in statistics; (c) the way in which a crime committed by more than one person is registered; (d) the application of the main offence rule (some countries regard simultaneous crimes considering only the main one—main offence rule—whereas others count each crime separately, which increases their rate of registered crimes); (e) the way in which multiple crimes are counted (whether it counts as one only crime or as multiple crimes if a victim reports the same crime more than once); and (f) the moment when data are gathered for crime statistics.

In view of this situation, many authors have stressed the need for establishing uniform definitions and for homologating data gathering systems (Savona and Stefanizzi 2007), in order to simplify transnational comparisons—essential for advancing in criminal investigation.

In fact, another relevant particularity to consider is the trans-nationalisation of organised crime. From a local point of view, this fact poses an enormous challenge in terms of measuring. If one group takes action in several countries at a time—as it frequently happens—the fact that one branch is identified in a specific location will only provide information about the volume corresponding to part of the criminal group, but not about the gross of it, seriously biasing measurement.

Another form of measuring consists of using the set of indicators for petty crimes as an indirect reference for presence and activity of organised crime, exploiting the potential of establishing some interrelation between both. Although theoretically it is true that it would be possible to find countries with high rates of conventional crimes and not so high rates of non-conventional crimes (organised crime), the most plausible reality is that a high rate of conventional crime corresponds to a high rate of organised crime (Van Dijk and Nevala 2002).

In fact, when high rates of conventional crimes are found paired with low rates of organised crime registered by the police, a higher real presence of organised crime is very likely. According to Van Dijk (2008), the number of cases of organised crime reported to the police is likely to be inversely proportional to the actual scale of the problem in a given country.

This statement has a relatively simple explanation: two of the factors that foster organised crime the most in a given scenario are: first the absence of specific control mechanisms to fight organised crime (lack of specialised police units, prosecutors, regulations, etc.) and second high levels of corruption (Albanese 2001). If these factors concur in a given geographical space, low rates of organised crime are likely to be found due to the low level of police detection (e.g. non-resolved homicides, a relatively low number of registered criminal organisations, etc.) (Giménez-Salinas Framis et al. 2009).

Statistics and police and penitentiary reports, along with those provided by the Administration of Justice, are not the only tools for measurement in the methodological aspect of organised crime analysis. On the one hand, there are victimisation surveys, which provide information about crimes suffered by a determined segment of the population (sample), usually regarding their high exposure to organised crime (risk of victimisation), and based on the population's faith in control instances. On the other hand, there are also self-reporting surveys (Roldán 2009); both typologies of tools have proven to be extremely useful for measuring conventional crimes (with samples of general population).

However, their use in the analysis of complex crimes poses difficulties (Medina Ariza 1999), which will be discussed hereafter. I would like to briefly mention the existence of other techniques—direct or participative observation and in-depth interviews, discarded a priori for not being ordinary and for the consequences they lead to.[1]

In most cases, the in-depth interview does not provide the knowledge of the structural and functional elements of criminal organisations, due to blatant difficulties regarding the access to the groups' upper or dominant layers, where the tendency to concealing activities is, logically, more intense the higher the decision-making level. Participant observation, that is, the participation of individual researchers in the internal dynamics of criminal groups, is simply impossible if the observer does not become at least an accessory or a witness of crime, unless the researcher is acting as a legal undercover agent at the same time. On the other hand, the participant observation, if achieved, raises ethical concerns about the legitimacy of putting scientific purposes before the fact of collaborating in criminal acts. Furthermore, researchers can be exposed to physical danger, and they could incur criminal responsibility.

Succinctly, the self-report consists of a free-participation and anonymous survey about the commission of lawless behaviour, aimed at perpetrators or potential perpetrators.

---

[1]Exceptionally, some researchers choose to employ these techniques, such as Dr. Wolfgang Herbert, who is one of the most eminent sociologists and experts in Japan of Austria, when he decided to infiltrate in the organised crime sphere of Osaka (Japan) to document his doctoral studies about japanese organised crime. Retrieved from Glenny M. (2008). Mc Mafia. El crimen sin fronteras. Barcelona: Destino Imago Mundi, p. 411.

Since it is based on perpetrators, the potential usefulness of this kind of instruments to measure organised crime is truly promising. Nevertheless, its weakness lies as much in the collaboration among the participants as in the reliability of their responses. Taking part in this kind of survey may be hindered in turn by realities such as coercion and threats that an organisation could exercise upon its members to preserve its security (intra-group violence). In addition, it is highly likely that those crimes eligible for being reflected in the self-report are not time-barred due to their severity (Kleemans and De Poot 2008).

Victimisation surveys have been somewhat successfully used to measure organised crime, on the grounds that they represent highly victimised groups (extortion and corruption). A noteworthy initiative in this field at the international level is represented by the International Criminal Business Survey (ICBS). The very same methodology used for the International Criminal Victimisation Survey (ICVS) was applied in 1995 to samples made up by executives of ten developed countries (Van Dijk and Terlouw 1996). Later, in 2000, a version of the ICBS questionnaire was used complemented with specific questions about organised crime and corruption.

At a national level, we must highlight the Encuesta Nacional de Victimización de Empresas ("National Survey of Companies Victimisation", ENVE in Spanish), carried out in 2012 by the Instituto Nacional de Estadísticas y Geografía ("Mexico's National Institute of Statistics and Geography", INEGI in Spanish). The effort made by INEGI in the light of Mexico's obvious difficulties over the last decade is very remarkable: a powerful and very active organised crime in some of the national territories, an open confrontation among various criminal organisations and between the latter and the government. The survey proved successful in terms of performance measures (despite having registered a black figure close to 88%, *black figure of criminality*, understood as the unknown crime rate and which, consequently, is not reflected in the statistics) which should be taken into consideration if we assume that it would be impossible to implement this kind of initiative in other criminal scenarios regarded as less dangerous owing to a lower incidence of criminal violence. This is due to the threats perpetuated by dominant criminal organisations to researchers and collaborators in order to preserve a situation of disinformation concerning their structures, activities and members.

Surveys about perception of organised crime have also been carried out within specific communities. On a similar line, the World Economic Forum included, from 2003 onwards, a survey to business people containing a question related to perception of organised crime in its annual Global Competitiveness Report.

At this point, measuring organised crime entails the recognition of the existing limitations to all methodologies involved in measuring the phenomenon of criminality by the criminological community. Consequently, it also has to be assumed that methodological complementarity is probably the most successful and accurate way (Fernández Villazala 2008) of approaching the study of organised crime.

5 Identification of Indicators and Weak Signals of Organised Crime: Particular Emphasis on Organised Group as Indicator and Last Trends

The starting premise that must be internalised as essential today is that security cut-off from intelligence is an outdated response. Therefore, identifying organised crime indicators and including them as optimal inputs for developing strategicwise criminal intelligence will allow us to delimit more accurately the main dimensions of the organised criminal phenomenon. This will enable a better evaluation of its capabilities, danger and insecurities.

The aim of developing trustworthy indicators lies in its ability to provide a contrasting empirical knowledge about the "organised crime" reality that facilitates an adjusted vision of it to the analyst when incorporated through the intelligence cycle. This vision will ultimately translate into useful intelligence (informed knowledge) to support the design of strategies against unlawful activities. A good product of strategic intelligence not only determines which is the current situation related to the phenomenon, but also provides explanations about the existence of that very phenomenon and sets likely evolutions or trends, defining possible and likely scenarios. It also enables the definition of objectives against organised crime and the establishment of policies and plans to implement and achieve the goals that have been set.

Following the main idea, the members of the Government Security structure shall abandon reactive random attacks and adopt a strategic planning perspective, in order to foster the impact of each operation against the organised crime group (Felbab-Brown 2013). This approach shall be strengthened by a quantitative criminological knowledge tool, properly processed and transformed into basic criminal intelligence. Furthermore, this approach shall constitute the basis for the design of focused dissuasion strategies and targeted action. The latter are considered as the most promising alternatives against organised crime (Garzón 2014).

Regarding organised groups as indicators, it is important to highlight something apparently obvious: organised crime is characterised by the existence of an organisation that directly conducts illegal activities. From this perspective, the main indicator to measure organised crime is the group of members that form the organisation (group). Therefore, a first approach to the evaluation of organised crime would consist of defining the number of active groups in the geographical areas analysed, as well as its characteristic features: territorial extension, type of organisational structure, major and secondary activities (i.e. whether multiple activities exist or not), if it is associated with other organisations and for how long, sophistication level, violence and corruption capacities, permeability towards state security bodies and presence in the financial and industrial sphere (partnerships with legal societies), among others.

The dimensions mentioned are especially relevant in order to have a holistic knowledge about the capacity for action, group resources, potential territories expansion, as well as its ability to influence society, economy and institutions.

All of them must be taken into account simultaneously for the evaluation of criminal organisations. Similarly, factors that contribute to the rise of economic, political and social crime are used as indicators for crime alert risk management (early alert) (Albanese 2001; Vander Beken 2004; Williams and Godson 2002). It should be pointed out that the specific statistics about criminal groups on which governments rely for organised crime risk assessments are based on very complex calculations (although not impossible) that seem impossible to verify and objectify (Carson 1984; Greenfield 1993; Blades and Roberts 2002; Van Duyne and Levi 2005).

On the other side, the adoption of criminal groups as indicators is controversial. Considering the number of groups as an indicator involves the inclusion of equivalent units in this category. Nevertheless, there are not equivalent units in organised crime. Taking the number of existing groups as the only criteria does not provide information about the extension of those groups, their typology, level of danger, level of institutional influence, abilities and resources.

Secondly, a group can develop different types of crime. Logically, not all the activities have the same importance when evaluating danger and consolidation level of the group. It is essential not to attach the same importance to the main activities, which constitute the group's business and illegal market, than to the instrumental illegal ones, related to the group's development, maintenance and survival (Abandinsky 2007).

According to the obstacles that have been mentioned, the evaluation and measuring of organised crime is extremely complex, since researchers must find new measuring options. In order to do so, they use indirect or context indicators (proxy) and organised crime perception measures. The indirect indicators represent a good alternative to measure the "immeasurable": unknown crime, organised crime and its emerging forms. Depending on the data used, the indicators can be pretty reliable (such as the number of stolen luxury motor vehicles that has not been found as an indirect indicator of the active presence of organised crime; since the vehicle insurance usually covers the incident, reporting is usually very high). However, indicators based on perceptions inevitably depend on the validity of such perception and definition of the people that are analysing the given case.

The concept of "weak signals" is abstracted from definitions of primary and secondary indicators, because the perception that, in reality, there is little tangible value to be extracted from isolated indicators, as there is potential for them, to be indicative of a variety of phenomena. However, when these indicators are grouped under certain conditions, such as proximity to a certain location and type of activity, they can begin to provide insights into to the presence or emergence of organised crime (CISC 2007). So, "weak signals of organised crime" may be defined as advanced indicators of change phenomena. They do not strike the potentially interested observer as such, but are complied on the basis of raw data. They are premature, incomplete, unstructured and fragmented informational material pointing to the emergence of challenging transformations. As advanced indicators that precede significant discrete one of events and/or novel developments in the rate and direction of trends, their analysis has the potential to facilitate the real-time alignment between organisational decision-making and changing external circumstances.

These predictors of future change pose problems of interpretation and represent a challenge to established models (Andrews et al. 2016). Thus, the practical significance of weak signal information is that it can be transformed into meaningful knowledge for to get knowledge about organised crime. In summary, "weak signals" are, for the purposes of this paper, considering them as "early warnings": advanced, incomplete and erratic symptoms of future problems derived from continuously evolving trends on organised crime.

The basic idea is that "certain indicators that were initially unsatisfactory can be joined together in order to create more reliable and valid indicators" (Aromaa and Heiskanen 2008). Nowadays, the integration of different data (from institutions, studies different context that are not directly related with the crime) represents one of the most promising options for criminology research.

Therefore, as it can be seen in the following example, low reporting rate of legal procedures related to organised crime can be explained by police corruption and political interference when processing and sentencing. Therefore, these low rates can actually mean a high rate in this type of crimes (Van Dijk 2007). Research conducted by the European Institute for Crime Prevention and Control of Helsinki (HEUNI), affiliated with the United Nations, follows the same direction. This organisation analyses the development of high rate crimes and the initiatives from the General Secretary of the United Nations related to emerging forms of crime (Malby 2012).

Professor Van Dijk, a very well-known researcher in this issue, has used interesting alternative measure instruments, such as the use of a composite index (Composite Organised Crime Index), based on several individual ones: organised crime perception index, unofficial economic index (most of these data come from the World Economic Forum surveys, investment risk evaluations of the Merchant International Group, World Bank Institute research and from official statistics), money laundering index, corruption index and, lastly, unresolved homicides index (Malby 2012). It is important to explain that the last index tries to determine the level of instrumental violence of those groups. This indirect form of measuring allows having a closer approach to the presence of organised crime in an area throughout the collateral and instrumental violence projected. This indicator correlates homicide rates with the presence of organised crime in a particular place. In other words, if there is a high rate of organised crime in an area, murder and violence reporting rates will be low, since corruption and impunity of these organisations enhances lack of governmental action, and therefore, all those crimes remain unsolved.

Some other possible and important approximation indicators are the non-related criminal activity indicator of criminal organisations or the structural complexity indicator (Moreno 2013). The Latin American Network of Security and Organised Crime (RELASEDOR, in Spanish, Red Latinoamericana de Seguridad y Delincuencia Organizada) is conducting an important task in crime analysis, and, more specifically, organised crime (as another reliable measure of indirect indicators). Some of their proposals are related to the number of missing people (forced disappearance) by organised groups, the number of deceased members of security forces in confrontation with members of criminal groups, number of murdered judges and prosecutors, number of murdered of threatened journalists, number of escaped prisoners controlled by these organisations, number of members from crime organisations that work at the security forces, number of terrestrial, aerial and maritime platforms used for drug transportation purposes, territorial controlled extension of the group (in square kilometres), number of intervened state institutions for criminal filtration (in local governments, for example) or the number of financial entities that have been also intervened, number of police interventions in those groups and money laundry indicators. All of them represent non-developed aspects for the investigation of organised crime. In general, they are all related to the use of violence, danger, impunity, logistic abilities, criminal filtration, resilience ability, impact on legal businesses, criminal activity volume and territorial control.

In summary, according to a fundamental syllogism that said "every crime activity leaves a trace" (Locard 2010), the next novel initiatives to be considered to know and about measuring organised crime are the EPOOLICE project (2015) that consist in a early pursuit against organised crime using environmental scanning, the Law and Intelligence systems; the analysis of the risk and the vulnerability factors of enterprises to be object racketeering by organised crime (Atanas Rusev et al. 2016); the detection, investigation and monitoring of organised crime groups using forensic intelligence: forensic data can contribute to the detection and follow-up of organised crime groups through a systematic approach (Sansó-Rubert Pascual 2015; Baechler 2016); the analysis of the links between elites and organised crime elaborated by the group of InSightCrime.org, to analyse the organised crime's impact on governance (2016); studies about organised crime and social media: how detecting and corroborating weak signals of human trafficking online (Andrews et al. 2016); the studies of criminal network structures (Campana 2016); or the analysis of how and why, the phenomenon of trans-nationalisation of organised crime is produced (Varese 2013; Sansó-Rubert Pascual 2016). Definitely, all these research programmes represent the most original and innovative initiatives in this area to date.

6 Conclusions and Future Challenges: Indicators, Strategies and Public Policies

The organised crime phenomenon will face some new dangers, but also new opportunities. This is why it is important to think about the future challenges towards security policies.

According to this approach, as the emerging forms of crime are improving, there must be an increased specialisation in instruments against organised crime. This specialisation requires the overcoming of classic paradigms, and it aims for innovative answers, both transversal and conciliatory (holistic). The simple policy and legal reaction will not work if it is not accompanied by social, economic, educative, environmental and legal initiatives.

At first, it seems that criminology studies are at a disadvantage. It must face that those with the ability to clarify the field of study (delinquents and, to a lesser extent, members of different organisations, services and other security institution forces) are not willing to do so, at least in a scientific way (Fernández Steinko 2013). Therefore, the problem finds its ignorance-based origin in the complexity of the object of study ("organised crime"), as with other sciences, but also in the existence of multiple academics interested in avoiding the problem. The issue gets worse because of the assumption of the organised crime as a non-empirical reality that can be simply measured with positivist criteria (Fernández Steinko 2013). As a social construction, it changes and evolves depending on social interests and perceptions in a particular social and political moment (Christie 2004; Aas 2007; Bauman 2002). This complicates the possible comparisons that could be made in the long run (if there is an important current criminal behaviour, tomorrow it could be the opposite). It also complicates all possible analysis based on comparisons between geographical areas, especially towards manifestations of organised crime. As an example, environmental organised crime (animal and plants trafficking, environmental pollution, fraudulent exploitation and natural and mineral resources trafficking) is very relevant and it is also monitored in Western Europe. Nevertheless, this crime is not classified as a criminal offence in other regions of the world. These circumstances, which should not frighten the researcher in order to get answers and which should foster the access to knowledge, must act as a stimulus to promote the engagement with scientific and criminological work. The rational identification of sensitive criminal behaviours and its measuring demands to deepen in precision and perfection analysis methods. Also, it requires the use of an analysis investigation technique based on transparent, open methodologies as compared with methodological regression and with lack of transparency of different sources.

The main objective is to elaborate public policies by those who are in charge and to determine the security strategies to have a reliable framework of organised crime as a consequence of a real diagnosis of causes and effects. As indicated previously, the purpose of the ensemble of indicators is to foster evaluation and measuring methodologies of organised crime. It is also used for facilitating detection and identification of any possible criminal success opportunity. In conclusion, it is aimed to promote the implementation of security preventive (and even proactive) initiatives based on the development of intelligence abilities. This formula contributes to the promotion and the strengthening of the democratic dimension of security through collection, systematisation, spreading and exchange of information and intelligence with previous awareness about what can be and what cannot be measured and under what terms, depending on the existing limitations.

References

Aas, K. F. (2007). *Globalization & crime*. Londres: Sage.
Abadinsky, H. (2004). *Drugs—An introduction* (5th ed.). Pacific Grove, CA: Brooks Cole.
Abandinsky, H. (2007). *Organized crime*. Belmont: Thomson Wadsworth.

Albanese, J. S. (2000). The causes of organized crime: Do criminals organized around opportunities for crime or do criminal opportunities create new offenders? *Journal of Contemporary Criminal Justice, 16*, 409–423.

Albanese, J. S. (2001). The prediction and control of organized crime: A risk assessment instrument for targeting law enforcement efforts. *Trends in Organized Crime, 6*(n° ¾), 4–29.

Albanese, J. S. (2008). Risk assessment in organized crime. Developing a market and product based model to determine threat levels. *Journal of Contemporary Criminal Justice, 24*, 263–273.

Albanese, J. S., & Das, D. K. (2003). Introduction: A framework for understanding. In J. S. Albanese, D. K. Das, & A. Verma (Eds.), *Organized crime: World perspectives* (pp. 1–17). New Jersey: Prentice Hall.

Alvazzi del Frate, A. (2008). Trends and methodological aspects in the International collection of crime and criminal justice statistics. In K. Aromaa & M. Heiskanen (Eds.), *Crime and criminal justice systems in Europe and North America 1995-2004* (n° 55, pp. 215–230). HEUNI publication series. Helsinki: Instituto Europeo de Prevención del Delito y Lucha contra la Delincuencia.

Andreas, P. (2011). Illicit globalisation: Myths, misconceptions, and historical lessons. *Political Science Quarterly, 126*(3), 1–23.

Andrews, S., & Brewster, B., & Day, T. (2016). Organised crime and social media: Detecting and corroborating weak signals of human trafficking online. In O. Haemmerlé, G. Stapleton, & C. Faron-Zucker (Eds.), *Graph-based representation and reasoning* (pp. 137–150). Lecture notes in computer science. Springer: Heidelberg.

Aromaa, K., & Heiskanen, M. (Eds.). (2008). *Crime and criminal justice systems in Europe and North America 1995-2004* (n° 55). HEUNI Publication Series. Helsinki: Instituto Europeo de Prevención del Delito y Lucha contra la Delincuencia.

Atanas Rusev, A., Garofalo, L., Giménez-Salinas, A., Jordá, C., & De Juan, M. (2016). Estorsione organizzata nell'Unione Europea: fattori di vulnerabilità. Center for the Study of Democracy.

Baechler, S. (2016). Detection, investigation and monitoring of organised crime groups using forensic intelligence: A promising path forward. Ecole des sciences criminelles, Université de Lausanne.

Bauman, Z. (2002). *La ambivalencia de la modernidad y otras conversaciones*. Barcelona: Paidós.

Bennett, R. R., & Lynch, J. P. (1990). Does a difference make a difference? Comparing cross-national crime indicators. *Criminology, 28*, 153–182.

Bernard, T., Snipes, J., & Gerould, L. (2010). *Vold's theoretical criminology*. New York: Oxford University Press.

Blades, D., & Roberts, D. (2002). Measuring the non-observed economy statistics. *OCDE, Brief* (n° 5). París: OCDE.

Campana, P. (2016). Explaining criminal networks: Strategies and potential pitfalls. *Methodological Innovations, 9*, 1–10.

Carson, C. S. (1984). The underground economy: An introduction (Measurement Methods). *Survey of Current Business, 64*, 21–37.

Christie, N. (2004). *A suitable amount of crime*. London: Routledge.

CISC Strategic Criminal Analytical Services. (2007). *Strategic early warning for criminal intelligence*. Technical report, Criminal Intelligence Service Canada (CISC).

De la Corte, L., & Giménez-Salinas, A. (2010). Crimen.org. Evolución y claves de la delincuencia organizada. Barcelona: Ariel.

Díez Ripollés, J. L. (2003). Política Criminal y Derecho Penal. Estudios. Valencia: Tirant Lo Blanc.

Edwards, A., & Gill, P. (2002). The politics of "transnational organized crime": Discourse, reflexivity and the narration of "threat". *The British Journal of Politics & International Relations, 4*(2), 245–270.

Felbab-Brown, V. (2013). *Disuasión focalizada, acción selectiva, tráfico de drogas y delincuencia organizada: conceptos y prácticas*. London: International Drug Policy Consortium.

Fernández Steinko, A. (2008). *Las pistas falsas del crimen organizado*. Catarata: Finanzas paralelas y orden internacional. Barcelona.

Fernández Steinko, A. (Ed.). (2013). *Delincuencia, finanzas y globalización*. Madrid: Centro de Investigaciones Sociológicas. Colección Academia.
Fernández Villazala, T. (2008). *La medición del delito en la seguridad pública*. Dykinson: Estudios de Criminología y Política Criminal. Madrid.
García Segura, C. (2015). La lucha contra la criminalidad transnacional organizada desde las Naciones Unidas y el rol de la Oficina de las Naciones Unidas contra la Droga y el Delito (UNODC): legitimidad e inmovilismo", en Ibáñez Muñoz, J. y Sánchez Avilés, C., (Dir.): Mercados ilegales y violencia armada (pp. 199–222). Los vínculos entre la criminalidad organizada y la conflictividad internacional, Madrid: Tecnos.
Garzón Vergara, J. C. (2014). Cómo responder al crimen organizado y terminar la guerra contra las drogas. Qué funciona, qué no funciona y cómo arreglarlo. Working Paper. Washington: Woodrow Wilson Center Update on The Americas.
Giménez-Salinas Framis, A. (2013). Impacto y consecuencias del crimen organizado: ¿quiénes son las víctimas? en Villacampa Estiarte, C. (Coord.). La delincuencia organizada: un reto a la política criminal actual (pp. 229–256). Navarra: Thomson Reuters. Aranzadi.
Giménez-Salinas Framis, A., Requena Espada, L., & de la Corte Ibáñez, L. (2009). La medición y evaluación de la criminalidad organizada en España: ¿Misión Imposible? en Revista electrónica de investigación criminológica (REIC), nº. 7.
Glenny, M. (2008). *Mc Mafia. El crimen sin fronteras*. Barcelona: Destino Imago Mundi.
Greenfield, H. I. (1993). *Invisible, outlawed, and untaxed: America's underground economy*. Westport: Praeger.
Howard, G., Newman, G., & Pridemore, W. (2000). Theory, method, and data in comparative criminology. In *Criminal justice 2000* (Vol. 4). Measurement and analysis of crime and justice. Washington, D.C.: United States Department of Justice, National Institute of Justice, pp. 139-211.
Ibáñez Muñoz, J., & Sánchez Avilés, C. (Dir.). (2015). Mercados ilegales y violencia armada. Los vínculos entre la criminalidad organizada y la conflictividad internacional, Madrid: Tecnos.
Kleemans, E. R., & De Poot, C. J. (2008). Criminal careers in organized crime and social opportunity structure. *European Journal of Criminology, 5*, 69–98.
Locard, E. (2010, reimpresión). Manual de técnica policiaca. Valladolid: Edición Facísimil, Editorial Maxtor.
Malby, S. (2012). Data collection on [new] forms and manifestations of crime. In M. Joutsen (Ed.), *New types of crime. Proceedings of the International Seminar Held in Connection with HEUNI's Thirtieth Anniversary*. HEUNI Publication Series, 74. Helsinki: Instituto Europeo de Prevención del Delito y Lucha contra la Delincuencia.
Medina Ariza, J. J. (1999). Una introducción al estudio criminológico del crimen organizado. en Ferré Olivé, J. C. y Anarte Borrallo, E. (Eds.). Delincuencia organizada: aspectos penales, procesales y criminológicos. Universidad de Huelva: Huelva.
Moore, R. H., & Fields C. B. (1996). Comparative criminal justice: Why study?. In C. B. Fields & R. H. Moore, Jr. (Eds.), *Comparative criminal justice: Traditional and non-traditional systems of law and control*. Prospect heights, Illinois: Waveland Press.
Moreno, F. (2013). Peligrosidad y daño directo del crimen organizado. en: Fernández Steinko, A. (Ed.), Delincuencia, finanzas y globalización (pp. 175–210). Madrid: Centro de Investigaciones Sociológicas. Colección Academia.
Morselli, C. (2009). *Inside criminal networks*. New York: Springer.
Mueller, G. O. W., & Adler, F. (1996). Globalization and criminal justice: A prologue. In C. B. Fields & R. H. Moore, Jr. (Eds.), Comparative criminal justice: Traditional and non-traditional systems of law and control. Prospect Heights, Illinois: Waveland Press.
Nelken, D. (1994). Whom can you trust? The future of comparative criminology. In D. Nelken (Ed.), *The futures of criminology* (pp. 223–225). London: Sage Publications.
Neuman, E. (1997). *Los que viven del delito y los otros. La delincuencia como industria* (2ª ed.). Argentina: Siglo veintiuno editores.
Nyer, J. S. (2004). La decadencia del poder blando de Estados Unidos. *Foreign Affairs, 83*(nº 2), 39–49.

Pulido Gragera, J., & Sansó-Rubert Pascual, D. (2014). A phenomenological analysis of terrorism and organized crime from a comparative criminological perspective. *Journal of Law and Criminal Justice, 2*(2). (American Research Institute for Policy Development. pp. 113–131).
Requena Espada, L. (2014). Principios generales de criminología del desarrollo y las carreras criminales. Barcelona: José María Bosch Editor, SA.
Requena Espada, L., Giménez-Salinas, A., & De Juan Espinoza, M. (2012). Estudiar la trata de personas. Problemas metodológicos y propuestas para su resolución, en Revista Electrónica de Ciencia Penal y Criminología, n° 14–13.
Resa Nestares, C. (1993). Crimen organizado transnacional: definición, causas y consecuencias. Accesible en http://www.uam.es/publicaciones. Universidad Autónoma de Madrid.
Reuter, P. (2001). Why does research have so little impact on American Drug Policy? *Addiction, 96*, 373–376.
Roldán Barbero, H. (2009). *Introducción a la investigación criminológica*. Estudios de Derecho Penal y Criminología. Editorial Comares: Granada.
Ronderos, J. G. (2005). The war on drugs and the military: The case of Colombia. In M. E. Beare (Ed.), *Critical reflections on transnational organized crime, money laundering and corruption*. Toronto: University of Toronto Press.
Ruiz-Funes García, M. (1960). *Criminología de la guerra; la guerra como crimen y causa del delito*. Buenos Aires: Editorial Bibliográfica Argentina.
Sansó-Rubert Pascual, D. (2008). Criminalidad organizada transnacional y seguridad internacional. En: Fernández Rodríguez, J. J., Jordán, J., y Sansó-Rubert Pascual, D. (Eds.), en Seguridad y Defensa hoy. Construyendo el futuro. Madrid: Plaza y Valdés Editores.
Sansó-Rubert Pascual, D. (2010). Organised crime and gender: Towards a redefinition of the role of women within criminal organisations? Revista de Criminología del Instituto Universitario de Investigación en Criminología y Ciencias Penales de la Universidad de Valencia (RECRIM), pp. 3–212. Accesible en http://www.uv.es/recrim
Sansó-Rubert Pascual, D. (2011). Globalización y delincuencia: el crimen organizado transnacional", en Jordán, J.; Pozo, P. y Baqués, J. (Eds.), La seguridad más allá del Estado. Madrid: Plaza y Valdés, pp. 135–157.
Sansó-Rubert Pascual, D. (2012). Estrategias de Seguridad, criminalidad organizada e inteligencia criminal: una apuesta de futuro. En J. J. Fernández Rodríguez, D. Sansó-Rubert Pascual, R. Monsalve, & J. Pulido Gragera (Eds.), en Cuestiones de Inteligencia en la sociedad contemporánea. Madrid: Ministerio de Defensa-Centro Nacional de Inteligencia, pp. 204–219.
Sansó-Rubert Pascual, D. (2015). Metodología y didáctica en el análisis y enseñanza de la criminalidad organizada como materia de estudio criminológico", en Revista Internacional de Investigación e Innovación en Didáctica de las Humanidades y las Ciencias, n° 2 (diciembre), pp. 111–135.
Sansó-Rubert Pascual, D. (2016). Nuevas tendencias de organización criminal y movilidad geográfica. Aproximación geopolítica en clave de inteligencia criminal", en Revista UNISCI/ UNISCI Journal, n° 41 (Mayo), pp. 181–203.
Sansó-Rubert Pascual, D., & Giménez-Salinas Framis, A. (2014). Crimen organizado. en De La Corte, L. y Blanco, J. M. (coords.): Seguridad nacional, amenazas y respuestas, Madrid: Editorial Lid, 2014, pp. 133–148.
Savona, E. U., & Stefanizzi, S. (2007). *Measuring human trafficking: Complexities and pitfalls*. New York: Springer.
Spalek, B. (2006). *Crime victims: Theory, policy and practice*. Basingstoke: Macmillan.
Sparrow, M. (2008). *The character of harms. Operational challenges in control*. Cambridge: Cambridge University Press.
Thoumi, F. (2005). The numbers game: Let's all guess the size of the illegal drug industry. *Journal of Drugs Issues, 35*(1), 185–200.
Van Dijk, J. (2007). Mafia markers: Assessing organized crime and its impact upon societies. *Trends in Organized Crime, 10*, 39–56.
Van Dijk, J. (2008). *The world of crime. Breaking the silence on problems of security, justice and development across the world*. Thousand Oaks: Sage.

Van Dijk, T. A. (2009). *Discurso y poder*. Barcelona: Gedisa.
Van Dijk, J., & Buscaglia, E. (2003). Controlling organized crime and corruption in the public sector. *Forum on Crime and Society, 3*, 1–2.
Van Dijk, J., & Nevala, S. (2002). Intercorrelations of crime: Results of an analysis of the correlations between indices of different types of conventional and non conventional crime. In P. Nieuwbeerta (Ed.), *Crime victimization in international perspective*. Den Haag: Boom Juridische Uitgevers.
Van Dijk, J., & Terlouw, G. T. (1996). "An International perspective of the business community as victims of fraud and crime. *Security Journal, 7*, 157–167.
Van Duyne, P., & Levi, M. (2005). *Drugs and money. Managing the drug trade and crime-money in Europe*. London: Routledge.
Vander Beken, T. (2004). Risky business: A risky based methodology to measure organized crime. *Crime Law and Social Change, 41*, 471–516.
Varese, F. (2013). *Mafias on the move: How organized crime conquers new territories*. New Jersey: Princeton University Press.
Von Hofer, H. (2000). Crime statistics as constructs: The case of Swedish rape statistics. *European Journal on Criminal Policy and Research, 8*, 78 y ss.
Von Lampe, K. (2004). Measuring organised crime a critique of current approaches .C. Petrus, Van Duyne, J. Matjaz, K. von Lampe, & J. L. Newell (Eds), *Threats and phantoms of organized crime, corruption and terrorism: Rhetoric and critical perspectives* (pp. 85–116). The Netherlands: Wolf Legal Publishers (WLP).
Walklate, S. (Ed.). (2007). *Handbook of victims and victimology*. Cullompton: Willan.
Walsh, J. M. (2004). Are we there yet? *Measuring progress in the U.S. war on drugs in Latin American. Drug war monitor*. Washington DC: Washington Office on Latin American.
Williams, P. (2005). Cooperación entre organizaciones criminales. en Berdal, M. y Serrano, M. (Comps.). Crimen transnacional organizado y seguridad internacional. Cambio y continuidad. México D. F.: Fondo de Cultura Económica, pp. 108–128.
Williams, P., & Godson, R. (2002). Anticipating organized and transnational crime. *Crime Law & Social Change, 37*, 311–355.

Part II
Methodology

Scanning of Open Data for Detection of Emerging Organized Crime Threats—The ePOOLICE Project

Raquel Pastor Pastor and Henrik Legind Larsen

1 Introduction

Open data provide an important source for detecting organized crime threats, emerging threats, as well as likely future threats. Such detection is depending on an ongoing monitoring and analysis of the relevant open data sources. For filtering out the potentially relevant information, we are in particular looking for development of *crime-relevant factors* (CRF) in the external environment (political, economic, social, etc.).

Due to the huge amount of open data and the limited human resources, this process needs to be supported by a system that can efficiently and effectively scan the open data sources and can filter and interpret the data into useful information for strategic analysis. An approach to such a system is provided by the ePOOLICE project (www.epoolice.eu) that developed and provided proof of concept of an advanced solution.

ePOOLICE is the acronym for *early Pursuit against Organized crime using environmental scanning, the Law and IntelligenCE systems*. It is a project co-funded by the European Union (EU) under its 7th Framework Programme for Research and Development (FP7). More precisely, it was granted as result of the 5th Security Research Call of this programme (FP7-SEC-2012-1) as an answer to a topic within the activity *Security and society* and within the area *Foresight, scenarios and security as an evolving concept*. The topic addressed by the ePOOLICE project was *developing an efficient and effective environmental scanning system as part of the early warning system for the detection of emerging organized crime*

R.P. Pastor
ISDEFE, Madrid, Spain
e-mail: rpastor@isdefe.es

H.L. Larsen (✉)
Department of Electronic Systems, Aalborg University, Aalborg, Denmark
e-mail: hll@es.aau.dk

© Springer International Publishing AG 2017
H.L. Larsen et al. (eds.), *Using Open Data to Detect Organized Crime Threats*, DOI 10.1007/978-3-319-52703-1_3

threats. Thus, the ePOOLICE project aimed at providing a "proof of concept" regarding the development of *an efficient and effective environmental scanning system* (the "ePOOLICE system") as part of *"the early warning system for the detection of emerging organized crime threats"* (providing the application context considered).

According to the EU request, the project objectives comprised:

- to conduct research into technological, actor-driven systems and tools which support environmental scanning to enable the rapid identification and qualification of new organized crime threats, by systematically monitoring the external environment for the detection of "weak signals" of upcoming opportunities and threats;
- to scan the environment to feed new and emerging threats into the serious and organized crime threat assessment processes;
- to identify a combination of technological resources and human actors that serves to improve the process of detecting and selecting new OC threats that warrant EU level analysis and EU-wide responses,

in order to improve

- the process of detecting and selecting new organized crime threats at EU level;
- the effectiveness of the end-users (LEAs, criminological institutes and private businesses);
- the strategic decision makers to counterbalance detected upcoming threats before they materialize;
- the understanding of the technologies and trends, leading to the strategic planning into security issues of all stakeholders.

As key aspects focussed as part of the project and the developed system prototype, can be highlighted:

- management of information and knowledge, including uncertainty management and provenance-based support of tracing findings back through the inference chain to the data sources;
- central information and knowledge repository to be used by the analysts and for "intelligent" (knowledge-based) scanning of open sources;
- information fusion for obtaining richer and more accurate information about entities and event of interest;
- methodology for application of knowledge-based environmental scanning for strategic analysis;
- analysis of legal, privacy and ethical issues in deploying such systems.

Though ePOOLICE—as an environmental scanning system for strategic early warning—is not focussed on collecting and organizing personal data, it may inadvertently extract such data that are common in open data and not easily identified due to the unstructured text form. Considering this, the ePOOLICE project developed a partial system solution using the *Privacy-by-Design* principles

(Cavoukian 2011). Though a full solution was not developed, the project provided a comprehensive analysis of the problem to better understand it (from legal, privacy and ethical point of views) and to provide a foundation for development of privacy preserving solutions (technical and organisational) as needed for deployment of such systems.

The remaining part of this chapter is organized as follows: In Sect. 2, we define the problem targeted by the ePOOLICE project. In Sect. 3, we present the principle architecture of the ePOOLICE system and some of its key components, emphasizing the central role of the Environmental Knowledge Repository. In Sect. 4, we introduce an important non-technical result of the project, namely the legal recommendation for development and deployment of ePOOLICE and similar system, regarding protection of personal data, ethics and privacy, considering relevant EU legislation.[1] In Sect. 5, we provide a brief overview over the project organization. We conclude in Sect. 6.

2 Environmental Scanning and Methodology

Since the development of an efficient and effective environmental scanning system was the main objective and subject of the ePOOLICE project, its definition and scope needed to be agreed for a common understanding among all parties within the project in order to get all efforts coordinated to a common and successful goal.

After some internal discussion, it was agreed to define the scope of the ePOOLICE project as follows, adapted from Conway (2009):

> The art of systematically exploring and interpreting the external environment to better understand the nature of trends and drivers of change and their likely future impact on Organised Crime

In their criminal (organized, serious) activity, organized crime groups (OCG) take advantage of changes in the environment, that is, of external factors in all dimensions of the human society, such as *political, economic, social and demographic, technological, legal* and *environmental*. These dimensions are also referred to as the *PESTLE dimensions* (from the first letter in the dimension name) and the factors as *PESTLE factors*. Among these factors, the factors of interest are the crime-relevant factors (CRF), comprising both crime-facilitating factors (CFF) and crime-inhibiting factors (CIF). In the study of factors driving future crime, the CFF are, of course, most interesting. The development of such factors is drivers of change which provides an uncertain, uncontrolled space that the OCG will try to take advantage of in the pursuit of their goal through criminal activities. This provides a potential threat to the society that needs to be aware of it in order to

---

[1] Chapter "Big Data—Fighting Organized Crime Threats while Preserving Privacy" provides a broader analysis and discussion of the underlying ethical and societal issues in deployment of environmental scanning systems like ePOOLICE.

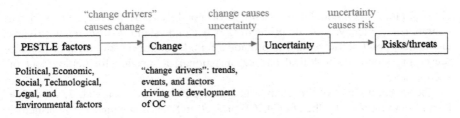

Fig. 1 Causal relationships from the PESTLE factors to the risks/threats

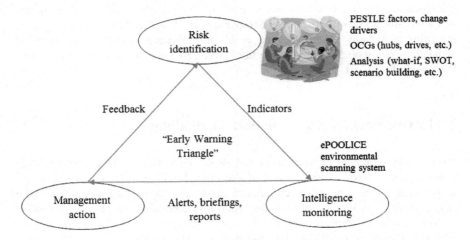

Fig. 2 Simplified view of the driving factors in the context of ePOOLICE

combat it through its political, legal and law enforcement instruments as needed to prevent the threat or to reduce its damage as much as possible without compromising on the fundamental human rights of the society. Figure 1 illustrates the causal relationships from the PESTLE factors to the risks/threats.

Figure 2 provides a simplified view of the driving factors in the context of the ePOOLICE system (for environmental scanning) and the methodological framework considered for applications of the ePOOLICE system, borrowing the Early Warning Triangle from Gilad (2003).

The EUROPOL report on *Exploring tomorrows organised crime* (EUROPOL 2015) characterizes the *key drivers for change and their impact on serious and organized crime* as follows:

> Organised crime will undergo profound and significant changes over the next decade in response to the availability of new technologies, changes in the environment such as economic challenges or developments in society and in response to law enforcement actions. Organised crime will undergo these changes whether or not experts agree on a new definition of organised crime and it is imperative for law enforcement to seriously consider the factors and driving forces that will shape serious and organised crime over the coming years.

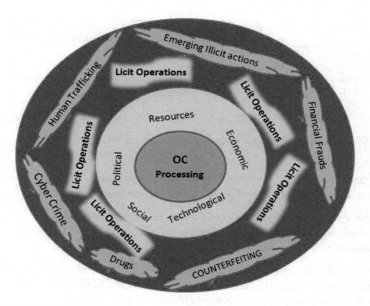

Fig. 3 PESTLE dimensions in the framework of organized crime activity

The environmental scanning fits in the strategy development and implementation cycle which consists of scanning, thinking, making decisions and planning. Each step is inter-dependent, but each needs to be considered as a separate process.

Scanning is the first step. It provides with information about what is happening in the external environment, about what is changing and about what there is a need to pay attention to on a continuing basis. This information has a focus on generating strategic options to consider. To identify these options, it is needed to get a good knowledge about external drivers of change.

This implies monitoring events that can be grouped in trends so that drivers, which moves trends in certain directions, can be detected and changes analysed.

Major changes in the environment induce changes in crime, so it becomes very important for law enforcement agencies (LEAs) to scan this environment, looking for its evolution in order to be best prepared for emerging organized crime threats.

Environmental scanning does not analyse known crime trends but non-criminal drivers of change and looks for their potential impact in criminal trends in the future. This is the reason to follow the PESTLE dimensions to assess key variables that define the factors of change; these dimensions in the framework of organized crime activity are illustrated in Fig. 3. Using a model such as PESTLE provides a starting point for the Environmental Scanning.

Accordingly, the ePOOLICE environmental scanning system provides a systematic overview of the surrounding environment to detect "weak signals" that may trigger a heavy and oriented situation awareness computation to better appreciate and anticipate emerging organized crime.

		OCG internal factors	
		Strengths	**Weaknesses**
External factors (PESTLE)	**Opportunities for OCG**		
	Threats to OCG		

Fig. 4 SWOT analysis scheme for analysing and identifying future OC threats

As depicted in Fig. 3, the environmental radar scans a virtual environment represented by huge volume of open-source information describing:

- tangible and intangible resources (water, petrol, climate, culture, …)
- political, legal and social organization
- technology advance and innovation

The described Environmental Scanning concepts link with Europol's definition in the EU Serious Organized Crime Threat Assessment, SOCTA 2013 (EUROPOL 2013) of *Crime Enablers* and *Crime-Relevant Factors* (CRF):

> Crime enablers are a collection of 'Crime-Relevant Factors' (CRF) that shape the nature, conduct and impact of serious and organised crime activities. CRF affect crime areas and the behaviour of both criminal actors and their victims. They include facilitating factors and vulnerabilities in society creating opportunities for crime or crime-fighting. They are the instruments by which serious and organised crime operates and are common to most areas and most groups.

> Certain enablers are particularly relevant for multiple crime areas and provide opportunities for different OCGs in their various activities. These horizontal crime enablers include the economic crisis, transportation and logistical hotspots, diaspora communities, corruption, legal business structures (LBS) and professional expertise, public attitudes and behaviour, risks and barriers of entry to criminal markets, the internet and e-commerce, legislation and cross-border opportunities, identity theft and document fraud.

> …

> CRF are facilitating factors and vulnerabilities in the environment that have an influence on current and future opportunities or barriers for OCGs[2] and SOC[3] areas. CRF are analysed via horizon scanning, which aims to identify future trends in society and future crime threats.

The SWOT (Strengths, Weaknesses, Opportunities and Threats) analysis method provides a supporting methodology for analysing how changes in the crime-relevant factors (in the PESTLE dimensions) may influence the development of organized crime (OC). This helps in identifying the OC threats and the possible

---

[2]Organized crime groups.
[3]Serious and organized crime.

scenarios through providing a structured approach to analyse how the OC groups (OCG) most likely will exploit the changes in these factors, as illustrated in Fig. 4.

3 The ePOOLICE Solution System

3.1 Overview

An important means for law enforcement in combatting organised crime is strategic early warning, which is heavily depending on an efficient and effective environmental scanning.

For this, the ePOOLICE project—in close collaboration with law enforcement partners, as well as criminological and legal experts—developed a prototype of an environmental scanning system implementing solutions applying the technological advances and breakthroughs as provided by the technological research partners. The solutions were tested and evaluated through running three realistic use case scenarios developed with the support of the end-user partners in the consortium. The main scenarios were in these OC areas: cocaine trafficking, trafficking of human beings (THB) and copper theft.

Central to the solution was the development of an Environmental Knowledge Repository (EKR) for storing and maintaining all relevant information and knowledge, including general and specific domain knowledge (ontologies), scanned information and derived, learned or hypothesized knowledge, as well as the metadata needed for credibility and confidence assessment, uncertainty management, traceability and privacy protection management. For effective and efficient utilization, as well as for interoperability, the repository applies a common representation form for all information and knowledge.

For effective and efficient scanning of the raw information sources, the project developed intelligent environmental radar that utilizes the knowledge repository for focusing the scanning.

The main solutions provided by the project are supporting:

1. *Monitoring the development of crime-relevant factors (CRF)*, in particular crime-facilitating factors (CFF), in the PESTLE dimensions of the environment.
2. *Detection of emerging organized crime*:
    a. Detect the existence of criminal activities typically run by organized crime.
    b. Discover organized crime and underlying criminal organizations as early as possible to prevent further formation of stronger, more resilient criminal systems.
3. *Prediction of the evolution of organized crime*. This requires environmental scanning system supporting analysing and developing scenarios of possible threats in the future.

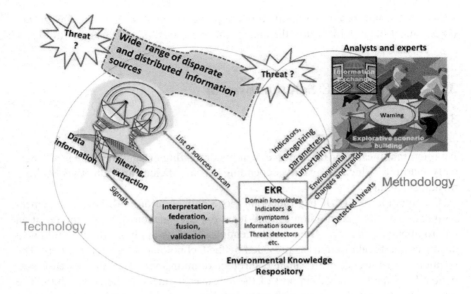

Fig. 5 ePOOLICE system idea with the human (analysts and experts) in the loop

The environmental scanning system (prototype) developed during the ePOOLICE project provides a systematic overview of the surrounding environment to better appreciate, assess and anticipate an emerging crime, by monitoring the environment and capturing in real-time relevant information present in heterogeneous open sources, including analysis reports, governmental information, web, news, academia, non-governmental and international organizations and subject matter experts. This is done in the methodological framework of the PESTLE dimensions of the environment and the indicators and signals to look for, as determined by the analysts and experts.

A key component of the framework is semantic filtering for identification of data items that constitute weak signals of emerging organized crime threats, exploiting fully the concept of crime hubs, crime indicators and facilitating factors, as understood by our user partners. The strategic analyst team interacts with the environmental scanning system according to the ePOOLICE methodology covering the whole technological-human system, i.e. with the user in the loop, as illustrated in Fig. 5.

The system supports:

- disparate information sources;
- multiple lingual support—with English and German demonstrated;
- dissemination and exchange of information and knowledge of potential interest to the law enforcement agencies;
- visualization of potentially emerging OC threats for OC threat assessment;

- early warning with alerts in cases of detection of potentially emerging new OC threats;
- storing and utilization of hypothesis and notes from user;
- utilization of user feedback on findings for refinement of the system's domain knowledge; analysis and decision making in addressing emerging OC threats, considering the validity and the seriousness of the detected threats.

The ePOOLICE methodology provides a structured approach to utilize the environmental scanning system efficiently and effectively with the user in the loop for monitoring heterogeneous information sources, as well as identifying and prioritizing indicators, as part of the strategic early warning process. For this purpose, the methodology includes the relevant legal, privacy and ethics aspects. The knowledge about a number of organized crime types, their facilitators, identifying signatures and indicators, etc., is part of the knowledge maintained in the Environmental Knowledge Repository (EKR).

The main functionalities available to the end-users (analysts) are:

- browse, navigate and zoom in the findings;
- select a geographical or/and temporal view for information of interest;
- refine the system's knowledge of OC types, criminal hubs, modus operandi, indicators, signals, etc.;
- refine the set of relevant sources, their importance and scanning frequency;

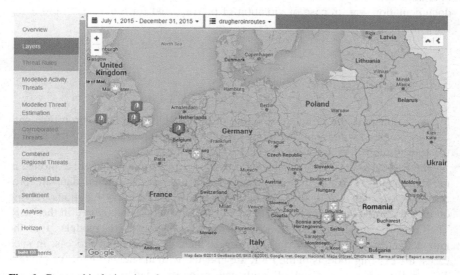

Fig. 6 Geographical view in a *heroin route* use case

- provide additional knowledge, information, hypothesis, hunches, etc., to be properly utilized by the system;
- share and discuss findings with relevant colleagues in the police collaboration;
- perform ad hoc queries and analyses in the EKR as needed in an analysis case.

Figure 6 shows an example of the geographical view in a use case applied for demonstration.

Further, the system can:

- use feedback from analysts to distillate and propose new indicators/signals and adjust its warning/alert levels;
- propose new sources to be scanned;
- evaluate and estimate the importance and optimal scanning frequency of sources, based on usage behaviour monitoring and possibly explicit user feedback and inputs;
- use information/knowledge while considering quality issues (e.g. completeness, accuracy, reliability) and the nature of the knowledge (e.g. factual, belief, hypothesis) for reasoning properly in answering a user query/question;
- monitor the EKR use for privacy issues.

All information and knowledge in the EKR is highly tagged (e.g. kind of knowledge, source, registration time, credibility, importance, sensibility) as needed for its proper utilization and for traceability of findings. All accesses by human operators and analysts, and by subsystems, are logged as required or needed for documentation, e.g. in case of violation of privacy, and for tuning and optimizing the system.

3.2 System Structure Overview

Figure 7 illustrates the principle structure of the ePOOLICE system prototype with the analyst user in the loop. The figure also shows the important role of the EKR as the central repository for all information and knowledge.

The EKR provides rich information, knowledge and models that are exploited in the data acquisition and the data structuring processes. Structured data are processed by the analysis, fusion, mining and knowledge discovery functions, relying on models stored in the EKR. The results of these processes are stored in the EKR. The result can further give rise to new or improved models for the knowledge-directed crawling and filtering in the data acquisition process; in such cases, the existing models in the EKR are updated with the new or improve models. The analyst users feed the EKR with new prior knowledge, new relevant information sources for data acquisition, hypotheses, etc.

Some of the key features of the system are described in the following subsections. Thus, Sect. 3.3 outlines the characteristics of the EKR. Section 3.4 shows

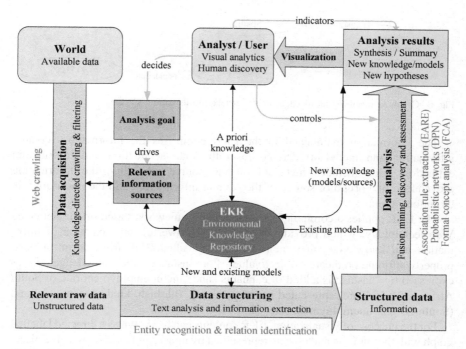

Fig. 7 Principle structure of the ePOOLICE system

examples of the visualization. Section 3.5 briefly introduces the data structuring (information extraction) process.

The ePOOLICE project further provided tools, as part of the ePOOLICE system prototype, to test and demonstrate the capability to exploit the information and knowledge gathered to predict organized crime threats. Section 3.6 briefly presents three of these tools.

3.3 The Environmental Knowledge Repository (EKR)

The Environmental Knowledge Repository (EKR) provides the central information and knowledge base for the ePOOLICE system.

The knowledge maintained in the EKR comprises common world knowledge and domain knowledge in the OC areas, as well as in the PESTLE domains with focus on crime-relevant factors (CRF) and indicators. It further comprises knowledge about relevant information sources (for data acquisition): their topic area, kind (news, reports, etc.), known reliability, access information (e.g. URL), etc. In the developed prototype for the "proof of concept", the domain knowledge is limited to that relevant to the selected use case scenarios (e.g. cocaine trafficking and human trafficking).

Fig. 8 Graph representing knowledge on Venezuela and its capital Caracas

The information maintained in the EKR comprises the information (content information and metadata) derived from the data acquisition and information extraction. This includes metadata on the source reliability, the information provenance (the origin or source of the content information) and the uncertainty in the information.

The EKR applies a common representational framework, based on the Semantic Web technology, which provides a simple, yet powerful representation of information and knowledge; in principle in the so-called RDF triple form: subject, property/attribute/predicate, object/value. For instance, the knowledge pieces expressed by "hashish is a kind of cannabis" and "the unemployment rate of South Africa is 26.6%" are represented by, respectively, (hashish, kind of, cannabis) and (South Africa, unemployment rate, 26.6%).

For the purpose of visualization, EKR contents can be shown as a directed labelled graph with the subjects and objects represented by nodes, and each knowledge piece, as modelled by an RDF triple, depicted by an edge (arrow) from the subject node to the object node. The edge represents the property/attribute/predicate element of the triple. Figure 8 shows graphs representing the three knowledge pieces (Venezuela, language, Spanish), (Venezuela, capital, Caracas) and (Caracas, total_population, 3,153,000), telling that Venezuela's natural language is Spanish and its capital is Caracas, which has a total population of 3,153,000 people.

The intended semantics of each element in a triple, and hence of the nodes and edges in the graph, is defined by a reference to the definition of the entity or concept. This is needed, since in many cases a name or term applied is, per se, ambiguous.

In ePOOLICE, the reference is provided by a Uniform Resource Identifier (URI) in the DBpedia.org knowledge. For instance, the triple (Venezuela, language, Spanish), as depicted by the upper part of Fig. 8, applies the following triple of references:

(<http://dbpedia.org/resource/Venezuela>,

<http://dbpedia.org/ontology/language>,

<http://dbpedia.org/resource/Spanish_language>).

Since the elements refer to semantic units, the knowledge representation is, in principle, language independent. Basic multilingualism is supported by the core RDF model by means of separating the semantic (language-independent) units and the language-dependent labels.

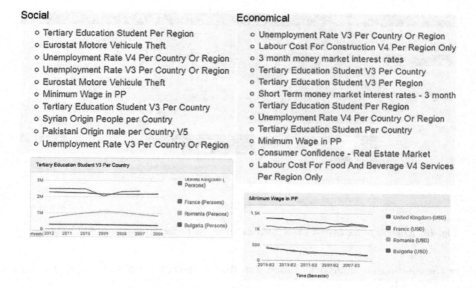

Fig. 9 Example of visualization of PESTLE CRFs

The main advantages of the EKR are:

- All relevant knowledge and information are stored and managed in a single highly secured base.
- The common representation form greatly facilitates integration of information and knowledge.
- Semantic, and, hence, language independent, knowledge representation.
- Domain knowledge (and information) is connected through the graph to semantically related knowledge, including common knowledge.

The research regarding the knowledge representation further led to proposal of a new scheme, "FrameBase", that uses linguistic frames to represent n-ary relations. While being consistent with the above-mentioned triple form, FrameBase provides flexibility and expressiveness for semantic integration of information from heterogeneous sources and facilitates the use of natural language processing techniques for information extraction and mining; see Rouces et al. (2015a, b, c).

3.4 Visualization

A GUI (Graphical User Interface) component was developed in the project as main visualization and dashboard interface of the ePOOLICE system. The GUI component features included:

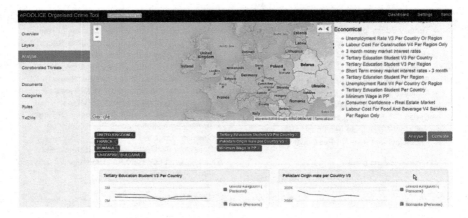

Fig. 10 Example of the analyst's user interface for analysing the development of PESTLE factors

- Session-based analytics, based on the organized crime scenarios used in ePOOLICE.
- Analytics and visualizations based on the corpus of documents and their extracted entities.
- Source document navigation and visualization.
- Map-based visualization of the Formal Concept Analysis threat concepts (see Sect. 3.6), including category-based filtering.
- Navigation and visualization of association rules provided by the EARE component (see Sect. 3.6).
- Map-based visualization of DPN statistical alarms (see Sect. 3.6).
- Analysis and correlation of OC indicators per region.

Figure 9 illustrates the visualization of CRFs in two PESTLE dimensions, as part of a user interface for the analyst.

The analyst can select geographical areas in a map and select the PESTLE dimensions and factors to be visualized, as illustrated in Fig. 10.

Section 3.6 shows screenshots of some of the tools developed to test and demonstrate the capability to exploit the information and knowledge gathered for analysis and prediction of organized crime threats.

3.5 Information Extraction

The information extraction comprises in particular recognition and extraction of entities and relations between entities. Figure 11 shows the flow in extraction of information from the crawled and acquired unstructured content (relevant raw data), typically in natural language text form. The extracted information, in the form of structured data, is stored in the EKR.

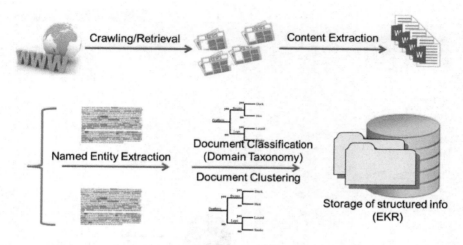

Fig. 11 Flow in information extraction from crawled and extracted content

Extraction of statistical information from, e.g. Eurostat, where data are typically in structured or tabular form, applies particular dedicated functions.

For published research from the ePOOLICE project on these techniques and related topics, see (Semmar et al. 2014, 2015; Wang et al. 2014).

A somewhat related research on selecting those concept/entity terms in a given semantic network knowledge representation that best summarize the contents of a group of phrases or sentences in a text is published in Ortiz-Arroyo (2013).

3.6 Information Fusion, Analysis and Threat Discovery

In the following, we briefly present some of the tools provided as part of the ePOOLICE system prototype to demonstrate and test the capability to exploit the information and knowledge gathered for analysis and discovery of likely threats, namely the EARE (Exception and Anomalous Rules Extractor) tool, the DPN (Distributed Perception Networks) tool and the FCA (Formal Concept Analysis) tool.

3.6.1 EARE (Exception and Anomalous Rules Extractor) Tool

The EARE uses association rule mining for discovery of potentially interesting relationships between entities. The relationships have the form of so-called association rules. There are three types of such rules:

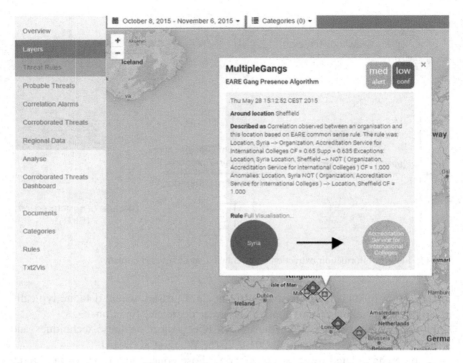

Fig. 12 Visualization of association rule-based threat indications in a use case

- Common sense—can indicate connections between entities
- Exceptions—can indicate exceptions to connections between entities
- Anomalous—can indicate alternatives to connections between entities

The following is an example of threat discovery through association analysis based on entities found in crawled open sources, in this case UK newspapers. The newspapers were crawled (data acquisition), and the entities were extracted (information extraction) and analysed by the EARE for entities that were often found together in a news article. The output was a list threat indications ("rules"), which were pinned on a map, based on the location found in the articles. For each of the pinned threats, the reason for the indication can be seen through clicking on the pin, as illustrated in Fig. 12.

Published research from the ePOOLICE project contains details about the techniques and related research; see Delgado et al. (2013), Ruiz et al. (2014a, b), (2015a, b, c).

Scanning of Open Data for Detection of Emerging ... 63

Fig. 13 Visualization of probable threats through DPN-based threat risk analysis in a use case

3.6.2 DPN (Distributed Perception Networks) Tool

The DPN applies a decentralized Bayesian fusion for detecting probable threats. It allows events and anomalies detection, as well as PESTLE factor estimation, since it provides a solid basis for the development of PESTLE process models correlating very different types of observable phenomena. The DPN can fuse heterogeneous data from various sources, e.g. statistical indicators, mining knowledge (from, e.g. EARE) and human experts.

The ePOOLICE system uses the DPN for detecting interesting events to trigger attention of human analysts and to focus the data mining and visual analytics processes to the relevant subsets of data. For instance, it can provide a threat level in a given "spatial area" based on the number of co-occurrence of entity "a certain organized crime group" and "spatial area" or act as a detector of emerging threat based on the same co-occurrence.

Figure 13 shows an example of discovery of probable threats through DPN-based threat risk analysis, which (as in the EARE example above) is based on entities found in UK news articles. The DPN produces a list of potential high-risk combination of entities, which are pinned on the map based on the location found in the documents. For each pin, the reason for the risk can be seen through clicking on the pin, as illustrated in the figure.

Fig. 14 Example of visualization of FCA analysis results

3.6.3 FCA (Formal Concept Analysis) Tool

Tools based on Formal Concept Analysis (FCA) (Wille 2005) were developed to analyse data concerning hypothetical and known organized crime threats. The FCA-based functionalities provided by these tools include: identifying groups of objects (entities) according to their similarity in terms of their attributes, describing connections (dependencies) between these groups and visualizing the groups and their connections by concept lattices. An example of the kind of applications of the FCA-based tools is knowledge discovery of the significant associations of indicators with situations in existing organized crime data. Figure 14 shows example (screen shot) of visualization of FCA analysis results.

Details about the use of FCA to detect and monitor organized crime and about related issues in knowledge management, etc., can be seen in published research from the ePOOLICE project (Andrews et al. 2013; Brewster et al. 2014a, b, c, d).

4 Legal Recommendations and Privacy Impact Assessment

Another major result of the ePOOLICE project is the release of legal recommendations for deploying systems like ePOOLICE. These recommendations were researched and developed by participating law specialists in close collaboration with other partners, in particular technological partners, during the whole project. Even if the ePOOLICE system is not aimed at collecting personal data, the open sources crawled often contain personal data, such as person names. Further, there is the risk that end-users of ePOOLICE systems may apply the system in other ways than intended which may involve collecting personal data. These are just few examples of the legal, ethical and privacy risks connected to the deployments of such, and indeed many other kinds of, information systems, in particular in connection to big data, social media and open-source intelligence. The legal and ethical analysis considers not only what is needed for demonstrating the technological proof of concept of the ePOOLICE system, but also the issues and feasibility for deployment of such systems.

The major findings from this research are presented in two reports. The first one is a 150 pages' public ePOOLICE report, *D3.3—LEGAL RECOMMENDATIONS* (De Marco et al. 2014). The second one is a 127 pages report, D3.4—FINAL REPORT (De Marco et al. 2015; Semmar et al. 2015), where a Privacy Impact Assessment methodology developed for the ePOOLICE project is applied to the ePOOLICE prototype. The executive summary (by E. De Marco, Inthemis) of the D3.3 report is reproduced below:

> The protection of an individual's right to private life, which also protects other freedoms either exercised in the private sphere (such as the freedom of expression) or that may be threatened by a privacy limitation (such as the right to non-discrimination), is ensured by the European Convention on Human Rights (E. Conv. H. R.) and the EU Charter on fundamental rights (EUCFR). The conditions laid down by these instruments to limit the right to private life apply to personal data protection. These conditions take the form of four principles (legal basis, legitimate purpose, necessity and proportionality), and each of these principles implies the respect of a set of requirements.
>
> In addition, several legal instruments, namely the EU Directives on personal data protection, the Council Framework Decision 2008/977/JHA, the Recommendation R. (87)15 of the Committee of Ministers of the Council of Europe and the Data Protection Convention, provide for more specific personal data protection. The last three of these instruments are applicable to processing operations for police purposes.
>
> The provision of these legal instruments may be seen, for the most part, as practical ways to ensure compliance with the E. Conv. H. R. and the EUCFR principles, within the framework of most data processing operations. This means that, on the first hand, these specific rules must be interpreted in the light of more general principles, including the principles of necessity and proportionality, and, on the other hand, that some processing operations may have to be based on additional specific legislation, to be compliant with the E. Conv. H. R. and the EUCFR (principle of legal basis). Such a specific legislation may have to implement additional appropriate safeguards, when generic ones are not sufficient or are not suitable for ensuring the legitimacy of the targeted processing operations (principles of necessity and proportionality).
>
> All the above-mentioned are legal principles, but also ethical ones, if we consider that the notion of "legal ethics" refers to the legislator's spirit, and to the value system and the philosophy underlying the legal system, and not only to the letter of the legal text. That leads to the interpretation of the concept of respect for private life (including the right to personal data protection) in a protective manner for the individual, taking into account the E. Conv. H. R. principles of legal basis, legitimate aim, necessity and proportionality.
>
> In addition, two ethical instruments have been developed and are progressively recognised by the legislator: privacy impact assessments and the privacy by design approach.

All the above-mentioned principles and instruments are applicable to the ePOOLICE research and to the ePOOLICE system with a view to its use by law enforcement authorities. Indeed, the collection of data on Internet public spaces may lead to collect personal data, even though such a collection might be unintentional.

Therefore, the current study proposes a methodology to be used to perform a privacy impact assessment (PIA) of the ePOOLICE project both within the framework of the research activities of the consortium and within the framework of the use of the system by law enforcement authorities, once the system is operational.

The methodology has been applied to the ePOOLICE research, and has led to the conclusion that a PIA is not necessary to assess the impact of the ePOOLICE research, since full compliance with legal and ethical requirements should be sufficient to ensure that privacy is appropriately protected, given the particularities of this project.

The first steps of the methodology have also been applied to the ePOOLICE project as it should be in its operational stage, and it has been concluded that a PIA is necessary to assess the impacts of the system to be used by law enforcement authorities, in addition to provide for functionalities and make recommendations to enable law enforcement authorities to ensure full compliance of their personal data processing with legal requirements.

To conclude, the current study provides several sets of recommendations for the ePOOLICE consortium, for the law enforcement authorities who will use the ePOOLICE system, and for EU Member States whose legislation will authorise this use. These two last sets of recommendations will be refined once the assessment of the ePOOLICE system to be used by law enforcement authorities will be completed.

The whole *D3.3—LEGAL RECOMMENDATIONS* report, is available at the ePOOLICE website (De Marco et al. 2014). As related to this topic, but with a particular focus on ethical and privacy issues, Chapter "Big Data—Fighting Organized Crime Threats while Preserving Privacy" of the current book provides a broad analysis and discussion of the underlying ethical and societal issues in deployment of environmental scanning systems like ePOOLICE. For other related published research in the ePOOLICE project, see Gerdes et al. (2013), Gerdes (2014, 2015a, b), Valls-Prieto (2013, 2014).

5 The ePOOLICE Project Organization

The project aimed to (cf. FP7 call) "conduct research into technologically-/actor-driven systems and tools which support environmental scanning". Accordingly, ePOOLICE was an end-user centric project that, as critical for its success, required a comprehensive understanding of the needs of potential end-users.

This understanding included the domain, the problem targeted, the user and stakeholder needs and requirements w.r.t. the solution system envisioned by the ePOOLICE project as detailed above. The development of the system prototype was guided by the end-user needs and the ethical/legal issues through active participations of LEAs and application of use case scenarios providing a

Fig. 15 ePOOLICE consortium

problem-oriented drive for the project, with the privacy-by-design approach to privacy security.

For this purpose, ePOOLICE benefitted all along the project life from the support by end-users and stakeholders (mainly LEAs but also criminological institutes), including the following consortium partners: EUROPOL, UNICRI (United Nations Interregional Crime and Justice Research Institute), Spanish Guardia Civil, Police and Crime Commissioner for West Yorkshire and the Bavarian Police Academy (Fachhochschule fur Offentliche Verwaltung und Rechtspflege in Bayern).

Further, an *End-User and Stakeholder Advisory Board* (EUSAB) was established at the very outset of the project to reinforce the involvement of the end-users (from different European security stakeholders, EU and Member State agencies and administrations), to ensure that the work conducted in the framework of the project was consistently of the highest quality and that the practical needs of end-users were being addressed and maintained throughout the duration of the ePOOLICE project.

All involved end-users, direct partners as well as associated partners, participated through the workshops that were organized along the project facilitating dialogue and collaboration among them and the research community in order to exchange the necessary information and know-how which drove the project from the beginning to successfully achieve its goals.

At the end of the project, a demonstration of the system prototype in various scenarios was carried out with the attendance of most of the involved end-users who found the system a useful tool for LEAs.

With a duration of three years, the ePOOLICE project was carried out by an international consortium composed of 17 partners from 7 different Member States of the European Union, led by the Spanish state-owned (Ministry of Defense) company *Ingeniería de Sistemas para la Defensa de España, S.A.*—ISDEFE.

The consortium was comprised by strong and complementary partners with different profiles and backgrounds—the aforementioned five end-users, academia, R&D organisations, SMEs and large companies—as it is shown in Fig. 15.

6 Conclusion

The project outcomes, of which some have been presented in the preceding sections, are the results of the joint efforts of all partners. In the presentation of the ePOOLICE system, several functionalities and components developed have not been mentioned due to space limitations of the chapter.

Several research results were presented at workshops/conferences and published in proceedings and scientific journals. A special session on the ePOOLICE topic "Environmental scanning for strategic early warning" was organized at FQAS 2013, the *10th International Conference on Flexible Query Answering Systems, Granada, 2013* (FQAS 2013; Larsen et al. 2013).

The project provided a proof of concept of an efficient and effective environmental scanning system as part of the early warning system for discovering emerging organized crime threats, according to the objectives presented in Sect. 1. The estimated Technology Readiness Level (TRL)[4] reached by the ePOOLICE components varied from 3 to 9, with an average of 6.2 and a median of 5.5.

Regarding the ePOOLICE system prototype developed and demonstrated, the main conclusions after the final review of the project can be summarized as:

- Overall, it can be concluded that the end-users' evaluation of the proposed environmental scanning system was positive.
- There was a great interest among end-users to further develop the system prototype into a useful product, given that the system demonstrated was found good and interesting, but not fit for usage in its delivered form.
- A possible further development towards a matured product should address the possibility to include together with the current ePOOLICE framework—environmental scanning for strategic use—also the support of investigative and operational use.
- For the proper inclusion of social media, especially for investigative use, a possible additional development must include further legal and ethical analysis, linked to the technological development. It may further include proposing a standard for privacy security in such systems, considering the results of this analysis.

---

[4]TRL 5: Component and/or breadboard validation in relevant environment, and TRL 6: System/subsystem model or prototype demonstration in a relevant environment. For a definition of the Technology Readiness Levels, see Table 2 of Appendix B in *Army Draft of TRLs for Software* (Graettinger et al. 2002).

Acknowledgements Research leading to these results has received funding from the European Union's Seventh Framework Programme (FP7/2007-2013) under grant agreement n° 312651.

References

Andrews, S., Akhgar, B., Yates, S. J., Stedmon, A. W., & Hirsch, L. (2013). Using formal concept analysis to detect and monitor organised crime. In H. L. Larsen et al. (Eds.), *Flexible Query Answering Systems (FQAS 2013), LNAI 8132* (pp. 124–133). Berlin, Heidelberg: Springer.

Brewster, B., Akhgar, B., Staniforth, A., Waddington, D., Andrews, S., Mitchell, S. J., et al. (2014a). Towards a model for the integration of knowledge management in law enforcement agencies. *International Journal of Electronic Security and Digital Forensics, 6*(1), 1–17. doi:10.1504/IJESDF.2014.060169

Brewster, B., Andrews, S., Polovina, S., Hirsch, L., & Akhgar, B. (2014c). Environmental scanning and knowledge representation for the detection of organised crime threats. In *Proceedings of Graph-Based Representation and Reasoning, 21st International Conference on Conceptual Structures, ICCS 2014, Iaşi, Romania, July 27–30, 2014, LNAI 8577* (pp. 275–280). doi:10.1007/978-3-319-08389-6_22

Brewster, B., Ingle, T., & Rankin, G. (2014b). Crawling open-source data for indicators of human trafficking. In *2014 IEEE/ACM 7th International Conference on Utility and Cloud Computing* (pp. 714–719).

Brewster, B., Polovina, S., Rankin, G., & Andrews, S. (2014d). Knowledge management and human trafficking: Using conceptual knowledge representation, text analytics and open-source data to combat organized crime. *ICCS,* 104–117. doi:10.1007/978-3-319-08389-6_10

Conway, M. (2009). Environmental scanning: What it is, how to do it.... *Thinking Futures, 2009.* https://socialenterprise.us/wp-content/uploads/2016/03/ES-Guide-April-09.pdf. Accessed October 19, 2016.

Cavoukian, A. (2011). *Privacy by design—The 7 foundational principles.* https://www.iab.org/wp-content/IAB-uploads/2011/03/fred_carter.pdf. Accessed October 10, 2016.

Delgado, M., Martín-Bautista, M. J., Ruiz, M. D., & Sánchez, D. (2013). Detecting anomalous and exceptional behaviour on credit data by means of association rules. In H. L. Larsen et al. (Eds.), *Flexible Query Answering Systems (FQAS 2013), LNAI 8132* (pp. 143–154). Berlin, Heidelberg: Springer. doi:10.1007/978-3-642-40769-7_13

De Marco, E., et al. (2014). *D3.3—Legal recommendations.* ePOOLICE report, Document ID: EPO-1408-WP03-001-V1.2-DV-RE. August 29, 2014. Available at https://www.epoolice.eu/. Accessed December 2, 2016.

De Marco, E., Semmar, N., et al. (2015). *D3.4—Final report.* ePOOLICE report, Document ID: EPO-1511-WP03-001-DV-RE. November 19, 2015. Available at https://www.epoolice.eu/. Accessed December 2, 2016.

EUROPOL. (2013). *SOCTA 2013—EU serious organized crime threat assessment.* Available from https://www.europol.europa.eu/content/eu-serious-and-organised-crime-threat-assessment-socta. Accessed October 10, 2016.

EUROPOL. (2015). *Exploring tomorrows organised crime. European Police Office, 2015.* https://www.europol.europa.eu/sites/default/files/Europol_OrgCrimeReport_web-final.pdf. Accessed October 10, 2016.

FQAS. (2013). *Proceedings of 10th International Conference on Flexible Query Answering Systems (FQAS 2013)* (Larsen et al., Eds.). Conference website: http://idbis.ugr.es/fqas2013/. Accessed October 10, 2016.

Gerdes, A. (2014). *A privacy preserving design framework in relation to an environmental scanning system for fighting organized crime.* Presented at the 11th IFIP TC9 Human Choice and Computers Conference—HCC 11, July 30th–August 1st, 2014, Turku Finland, ICT and

Society. Published in: *ICT and Society, IFIP Advances in Information and Communication Technology* (Vol. 431, pp. 226–238). doi:10.1007/978-3-662-44208-1_19

Gerdes, A. (2015a). State of the art—An ongoing EU research project: Security technology facilitating LEAS long term strategic decision making: ePOOLICE. In *Workshop on Technology Led Policing—Balancing Crime Fighting and Citizens' Privacy, January 14–16, 2015, at Council for scientific and industrial research (CSIR), Pretoria, South Africa.*

Gerdes, A. (2015b). ePOOLICE security technology: Fighting organized crime whilst balancing privacy and national security. In *10th International Conference on Cyber Warfare and Security ICCWS-2015, South Africa, March 24–25* (pp. 102–107).

Gerdes, A., Larsen, H. L., & Rouces, J. (2013). Issues of security and informational privacy in relation to an environmental scanning system for fighting organized crime. In H. L. Larsen et al. (Eds.), *Flexible Query Answering Systems (FQAS 2013), LNAI 8132* (pp. 155–163). Berlin, Heidelberg: Springer. doi:10.1007/978-3-642-40769-7_14

Gilad, B. (2003). *Early Warning—Using Competitive Intelligence to Anticipate Market Shifts, Control Risk, and Create Powerful Strategies* (AMACOM, 2003).

Graettinger, C. P., Garcia, S., Mel, C., Peter, S., Rdec, C., Siviy, J., et al. (2002). *Using the Technology Readiness Levels Scale to Support Technology Management in the DoD's ATD/STO Environments.* SPECIAL REPORT, CMU/SEI-2002-SR-027.

Larsen, H. L., Martin-Bautista, M. J., Vila, M. A., Andreasen, T., & Christiansen, H. (Eds.). (2013). Flexible query answering systems (FQAS 2013). In *Proceeding from the 10th International Conference, Granada, Spain, September 18–20, 2013, LNAI 8132.* Berlin, Heidelberg: Springer (With Special Session devoted to ePOOLICE: Environmental scanning for strategic early warning). doi:10.1007/978-3-642-40769-7

Ortiz-Arroyo, D. (2013). Analysis of semantic networks using complex networks concepts. In H. L. Larsen et al. (Eds.), *Flexible Query Answering Systems (FQAS 2013), LNAI 8132* (pp. 134–142). Berlin, Heidelberg: Springer.

Rouces, J., Melo, G. D., & Hose, K. (2015a). FrameBase: Representing N-Ary relations using semantic frames. In *12th Extended Semantic Web Conference (ESWC 2015).* doi:10.1007/978-3-319-18818-8_31

Rouces, J., Melo, G. D., & Hose, K. (2015b). Integrating heterogeneous knowledge with framebase. *Semantic Web–Interoperability, Usability, Applicability an IOS Press Journal.* http://www.semantic-web-journal.net/system/files/swj1239.pdf. Accessed October 19, 2016.

Rouces, J., Melo, G. D., & Hose, K. (2015c). Representing specialized events with FrameBase. In *4th International Workshop on Detection, Representation, and Exploitation of Events in the Semantic Web (DeRiVE 2015)* (pp. 58–69).

Ruiz, M. D., Gómez-Romero, J., Martín-Bautista, M. J., Sánchez, D., & Delgado, M. (2014b). Meta-association rules for fusing regular association rules from different databases. *FUSION 2014.*

Ruiz, M. D., Gómez-Romero, J., Martín-Bautista, M. J., Sánchez, D., & Delgado, M. (2015a). Fuzzy meta-association rules for information fusion. *FUSION 2015.*

Ruiz, M. D., Gómez-Romero, J., Martín-Bautista, M. J., Sánchez, D., Vila, M. A., & Delgado, M. (2015b). Fuzzy meta-association rules. *IFSA-EUSFLAT.*

Ruiz, M. D., Martín-Bautista, M. J., Sánchez, D., & Delgado, M. (2015c). Discovering fuzzy exception and anomalous rules. *International Journal of IEEE Transactions on Fuzzy Systems, 24*(4), 930–944. doi:10.1109/TFUZZ.2015.2489240

Ruiz, M. D., Martín-Bautista, M. J., Sánchez, D., Vila, M. A., & Delgado, M. (2014a). Anomaly detection using fuzzy association rules. *International Journal of Electronic Security and Digital Forensics, 6*(1), 25–37. doi:10.1504/IJESDF.2014.060171

Semmar, N., Zennaki, O., & Laib, M. (2014). Using cross-language information retrieval and statistical language modelling in example-based machine translation. In *36th Conference on Translating and the Computer* (pp. 36–44). http://www.tradulex.com/varia/TC36-london2014.pdf#page=37. Accessed October 19, 2016.

Semmar, N., Zennaki, O., & Laib, M. (2015). Evaluating the impact of using a domain-specific bilingual lexicon on the performance of a hybrid machine translation approach. In *Proceedings of Recent Advances in Natural Language Processing, Hissar, Bulgaria.* https://www.aclweb.org/anthology/R/R15/R15-1075.pdf. Accessed October 19, 2016.

Valls-Prieto, J. (2013). Legal and ethical problems linked to the processing of personal data for the purposes of prevention, investigation, detection or prosecution of criminal offences, Detection or prosecution of criminal offence. In *ITC Law 2013* (pp. 67–75). Department of Technology and Law, Porto University.

Valls-Prieto, J. (2014). Fighting cybercrime and protecting privacy: DDoS, spy software, and online attacks. In *Handbook of research on digital crime, cyberspace security, and information assurance* (pp. 146–155). doi:10.4018/978-1-4666-6324-4.ch010

Wang, W., Besançon, R., Ferret, O., & Grau, B. (2014). Semantic clustering of relations between named entities. In: *Advances in natural language processing, 8686.* doi:10.1007/978-3-319-10888-9_36

Wille, R. (2005). Formal concept analysis as mathematical theory of concepts and concept hierarchies. In B. Ganter et al. (Eds.), *Formal Concept Analysis, LNAI 3626* (pp. 1–33). Berlin, Heidelberg: Springer.

Foresight and the Future of Crime: Advancing Environmental Scanning Approaches

Martin Kruse and Adam D.M. Svendsen

1 Introduction

This chapter is substantially based on the transnational organised crime-related open source intelligence (OSINT) modelling, Strategic Futures, and more generally ranging futures studies work that we—at the Copenhagen Institute for Futures Studies (CIFS), based in central Copenhagen, Denmark—undertook as part of our contribution to the three-year European Union (EU) Framework Programme Seven (FP7) 'ePOOLICE' project, during 2012–2015. We are indebted to our consortium partners and to the law enforcement end-users for their helpful feedback at numerous workshops held across Europe during the course of the project.[1]

According to the European Union research in foresight, "foresight in the broad sense—often called **prospective** or **forward looking activities**—aims to shed light on different options for the future that may encompass different pathways of social and/or technological developments".[2] Foresight thus is the practice of analysing how changes that might happen in the future may affect organisational structures, policies, value chains, etc., and on the basis of this anticipative manoeuvre adapt accordingly. People acknowledge that with time, technology and other developments change the world we live in, yet when preparing for the future forecasters have a tendency to make more cautious, even conservative, forecasts based on mainstream present development trajectories resulting in a 'more of the same'-type of scenario.[3]

---

[1] Publicly available details of the project can be found via its website of www.epoolice.eu.
[2] EU Commission (2014).
[3] See, for instance, Dixon (2016) see also Andrew (2004); Svendsen (2012a, b), p. 103.

M. Kruse · A.D.M. Svendsen (✉)
Copenhagen Institute for Futures Studies, Copenhagen, Denmark
e-mail: adam@asgonline.co.uk

M. Kruse
e-mail: mkr@iff.dk

© Springer International Publishing AG 2017
H.L. Larsen et al. (eds.), *Using Open Data to Detect Organized Crime Threats*, DOI 10.1007/978-3-319-52703-1_4

The definition of foresight presented above helps frame much of what follows in this chapter, with more of an emphasis on adopting 'environmental scanning' approaches over merely 'prediction', as the latter approach has more reductionist single-point connotations than 'environmental scanning' approaches that have the added value and benefit of boasting more multiple futures applicability.

Foresight investigates change. The world changes as new technologies are invented and new business models appear as a response to the changes. Crime is no different. Organised Crime Groups (OCG) are in many aspects like any other business, and will behave according to rational business needs and are hence governed by market dynamics, enterprise, and the pursuit of profit. They can then be analysed using conventional economic models. When new technologies appear that can aid criminals in their pursuit of profit, they will use them. If new technologies appear that helps law enforcement crack down on criminals, OCGs will try to counter this new threat to their enterprise by adapting and using defensive tactics and maybe using other new technologies to counter the evolving threat from law enforcement. The use of foresight can thus be used in relation to crime in the same ways foresight is used in many other areas.

Substantially, the work covered in this chapter relates to CIFS' contribution towards the creation of a refined threat assessment methodology relating to OCGs. This work had the overall aim of supporting the development of an efficient and effective environmental scanning system, as part of an early warning and horizon-scanning mechanism with the primary purpose of providing foresight relevant to future crime and insights into its closely associated trends.

This generic approach adopted towards the challenges at hand includes ranging both across and into areas of international or 'globalised strategic risk' (GSR) and multi-functional to special operations (MFOs-SO/SpecOps) contexts, capturing the highly complex nature of the strategic landscapes to operational terrains encountered contemporaneously and anticipated into the future. Collectively, those cover, inter alia, the tasks of: crisis management, peacekeeping and humanitarian operations, cyber, counter-insurgency, counter-terrorism, counter-proliferation, the countering of transnational organised crimes, and so forth.[4]

Moreover, a brief literature survey suggests that there already exists substantial noteworthy research on several of the foresight and future crime themes covered throughout this chapter. Relevant areas, such as the regularly referenced 'intelligence-led policing' (in summary 'seen as a process model with an organisational infrastructure to support how policing in general should be conducted'[5]) and the concept of 'predictive policing' (essentially defined as involving: 'the desire to use [analytical] technology to predict [using statistics] places for police to pay attention'[6]) are becoming more widespread, albeit in somewhat differing guises in

---

[4]See as characterised in Svendsen (2015a, b), p. 58.
[5]Ratcliffe (2016 [2ed.]), p. 64.
[6]Ibid., p. 3.

their more precise details.[7] The findings of this chapter effectively build further on those earlier insights.[8]

In its main, this chapter consists of three parts. It begins its analysis by (i) examining the 'environmental scanning' dimensions of foresight relating to the future of crime by beginning with an overview of environmental scanning; before, secondly, (ii) examining a series of models relevant to undertaking environmental scanning. Finally, third, (iii) the chapter finishes with some overall conclusions, which bring together several of the foresight and future crime-related insights examined throughout. Some illustrative figures are presented alongside the text to assist with explanation relating to these continuing to be highly complex areas of endeavour.

2 Environmental Scanning: An Overview

Engaging in environmental scanning stands out as a useful place to begin. The concepts of environmental scanning, early warning and foresight arise out of the need for wanting to know in advance what will happen in the future. This is with some degree of probability, in order to react ahead of time, and thus reduce potentially adverse impacts for an organisation or a government.

The concept of environmental scanning is ill-defined and could in principle be used about any activity that scans for information in order to better understand how to respond to the changing environment. Thus, in principle, meteorological forecasting is an example of environmental scanning, as well as citing the use of opinion polls that inform politicians about the mood of their constituents or 'the people' (i.e. the public, civil society, etc.).

The concept, however, has been used mostly in business strategy and as a policy tool that systematically gathers a broad range of information about emerging issues and trends in an organisation's political, economic, social, technological, or environmental (including ecological) sectors. This introduces the concept of the PESTLE model, which is discussed further in this chapter. See also Svendsen (2017), p. 72, p. 86.

As already introduced above, the idea of environmental scanning is closely related to the concept of 'horizon scanning' and together with 'weak signals' and 'early warning' they frequently lack conceptual clarity. The term 'horizon scanning' concerns the activity of 'looking ahead'. Things on the horizon lack clarity, with greater degrees of uncertainty being involved. The term horizon scanning thus encompasses many of the attributes related to 'environmental scanning', 'early warning' and 'weak signals'.

---

[7]Perry et al. (2013) for further recent insights see also discussion in Ratcliffe (2016), p. 220, col. 1.
[8]While much work can be cited here, this literature notably includes, e.g., Ansoff (1975), pp. 21–33; Martinet (2010), pp. 1485–1487; Schoemaker et al. (2013), pp. 815–824; Lockwood (2013); see also the fifty-five structured analytic techniques as detailed in Heuer and Pherson (2014 [2ed.]).

The concept of 'weak signals' forms a useful jumping off point for our analysis. The idea of 'weak signals' was first introduced by H. Igor Ansoff in his famous paper on strategy: *'Managing Strategic Surprise by Response to Weak Signals'*, published in 1975. Ansoff wrote:

> We might call this graduated response through amplification and response to weak signals, in contrast to conventional strategic planning that depends on strong signals. Such a practical method for planning a graduated response can be developed. The first task is to explore the range of weak signals that can be typically expected from a strategic discontinuity.[9]

H. Igor Ansoff defines weak signals as signals that have a higher degree of uncertainty in part because of them pointing in inconsistent directions. Without using the term 'early warning', Ansoff implicitly works with the concept, which can be seen amongst others in his use of a quote by US President Abraham Lincoln from 1858:

> If we could first know where we are and whither we are tending, we could better judge what to do and how to do it.[10]

The concept of 'early warning' is also implicitly presented in the methodology Ansoff describes when dealing with weak signals, which can be found in his description of the dynamics of response. The aim of weak signals is to start [a] response much earlier and finish earlier,' through 'utilising [the] weak signals'.[11]

Another aspect of Ansoff's weak signal concept is important to note. Weak signals forbear a strategic discontinuity—for example, as Ansoff has noted: 'The first task is to explore the range of weak signals that can be typically expected from a strategic discontinuity.'[12] Thus the weak signal is about information that departs from an expected *modus operandi* (or equally anticipated *modi operandi*—in their plurality).

Weak signals revolve around discontinuities and are therefore representative of anomalies. However, anomalies are not necessarily discontinuities. While these characteristics of the weak signal are part of Ansoff's methodology and are also found in subsequent literature, these characteristics are not (so) prevalent in early warning systems that seek to protect infrastructure and guard against physical hazards. Here, the concept of early warning is used more to refer to the premature nature and timeliness of the warning than on the estimated impact or discontinuity, and the concept of weak signal describes imperfect information often obfuscated by confounding factors, which, in turn, are frequently characterised as 'noise'.[13] Long-term prediction horizons will by nature blur the indicators' signalling power and thus come to resemble a weak signal.

---

[9]Ansoff (1975), pp. 21–33.
[10]Ibid.
[11]Ibid.
[12]Ibid., p. 23.
[13]See for recent valuable insights, e.g., Silver (2012).

Table 1 Confusion matrix

	Crisis	No crisis
Signal is issued	TP true positive	FP false positive
Signal is not issued	FN false negative	TN true negative

Table 2 Derivation from the confusion matrix

The true positive rate (TPR) is the fraction of correctly predicted crises	TP/(TP + FN)
The 1-TPR is denoted as the Type I error rate, false alarm which represents the fraction of missed crises	FN/(TP + FN)
The noise or false positive ratio (FPR) also denoted the Type II error rate represents the fraction of false alarms, i.e. signals wrongly issued	FP/(FP + TN)
Predictive power is defined by the noise to signal ratio	TPR/FPR

However, many early warning systems (EWS) are crisis warning systems that monitor the development of pertinent variables. Statistically based early warning systems (EWS) employ a specific set of leading indicators with established predictive power. The statistical evaluation of potential early warning leading indicators requires examining variables that historically have signalled the build-up of risk and impending crisis often using a multi-variables approach to reduce the number of false alarms and improve the overall performance of the system (Tables 1 and 2).

From the ratios introduced above, the systems can be developed in such a way that an optimal trade-off between missing crises (Type I error) and issuing false alarms (Type II error) is found.

Ansoff does not distinguish between risk and uncertainty. Risk can be defined as a probability of an event, threat, and/or damage which may occur and which can be predicted on the basis of a distribution of previous events or trends. Thus, risk = likelihood × impact. An uncertainty can be defined as an unexpected future event or outcome of a future event that cannot be predicted often because of their rare and seemingly random distribution. It goes for both risk and uncertainties that they are often used about negative events, but can also be used about positive events—e.g. positive risks, possibilities and opportunities.

Ansoff's definition of the weak signal leans towards the latter definition, but in the context of foresight in relation to the future of crime, focus covers both the disruptive properties of the weak signal and the imperfection of information that defines the weak signal. There remains scope for the further adoption of weak signal utilisation, particularly in overall and extracted 'out of' background 'noise' contexts.[14]

In the next section of this chapter, we go further reviewing contemporary models focused on evaluating organised crime and realising strategic early warning (SEW). We also see that, as with the plethora of organised crime strategies and definitions

---

[14]Ansoff (1975) pp. 21–33.

that exist, there is no single dominant, universally established, or indeed overriding model for evaluating organised crime. This condition presents some obvious difficulties, such as relating to making targeting processes somewhat challenging.

However, instead of being taken as more of a negative trait, this lack of conceptual rigidity surrounding the evaluation of 'organised crime' in fact allows for some valuable scope and room for manoeuvre, both strategically and operationally. This readily malleable or 'soft' definition of organised crime enables at least a degree of greater operational flexibility and agility. These features are particularly useful during subsequent analysis and assessment processes, helping practitioners to answer as comprehensively as possible 'what is it?' questions and 'what does it mean?' queries, which continuously surface during analysis and assessment/estimate processes. These considerations appear across many different time and space horizons when dealing with wide-ranging and highly complex organised crime issues including, for example, their impact.[15]

At their most concentrated, customer/end-user requirements can be distilled down to having clear overlaps with the distinctly definable 'benchmarking' or 'best practice standards' criteria of 'STARC'—STARC consisting of 'Specificity, Timeliness, Accuracy, Relevance and Clarity' qualities.[16] If these customer/end-user requirements are not met, at least in terms of most of them or sufficiently substantially across their broad-ranging spectrum, models-to-strategies will increasingly falter.

Offering some preliminary conclusions, as recent RAND research has found: '... there is ambiguity in the definitions and analysis procedures adopted by police agencies to assess risk in the case of organised crime and...', as they go on to warn, '*this is a serious weakness.*'[17] Ultimately, as discussed elsewhere, 'Defining organised crime is problematic because of the vast range of crime activities that require some sort of collective organisation'. Continuing, 'the term ... is loosely applied to amorphous groups of offenders committing offence types that are believed to require some sort of organisation, as well as to mafia or triad-type syndicates', which maybe short- to long-term enduring.[18] Further strengthening can be undertaken into the future, as this chapter goes on to demonstrate via further breaking down the challenges encountered in a highly functional-intended manner.

---

[15]See also, e.g., Svendsen (2012a), p. 143, 150; see also references to 'soft law' and 'low levels of legalisation' concepts in sources, e.g., Jojarth (2009), esp. p. 14 and 272; see also '2.8—Soft law' in Dixon (2013 [7ed.]), p. 52.
[16]See these criteria as given in Hall and Citrenbaum (2012), p. 81.
[17]Perry et al. (2013).
[18]Ratcliffe (2016), p. 41.

3 Models for Environmental Scanning Relating to Future Crime

3.1 Background Context—Where We Are Currently

Contemporary mechanisms for foresight relating to the future of crime are now evaluated. As a first example, the Serious and Organised Crime Threat Assessment (SOCTA) is a strategic report developed by the European Police Office (EUROPOL) and is based on information retrieved from law enforcement agencies located across Europe. It identifies and assesses threats to the EU and analyses vulnerabilities and opportunities for crime and is thus a future-oriented threat assessment tool. As the SOCTA states:

The aim of the SOCTA is to:

- analyse the character or threatening features of organised crime groups (OCGs)
- analyse the threatening features of serious and organised crime areas of activity (SOC areas)
- analyse threatening aspects of OCG and SOC areas by region
- define the most threatening OCGs, criminal areas and their regional dimension[19]

An example from Canada, however, provides more in-depth (unclassified) insight into the methodological issues encountered in strategic early warning (SEW) work. Offering valuable insight, the Canadian example is now examined further. In its *'Strategic Early Warning for Criminal Intelligence: Theoretical Framework and Sentinel Methodology'* document (2007), the Criminal Intelligence Service Canada (CISC) provides several useful insights into: (i) a practical methodology for systemically understanding and evaluating organised crime issues; and (ii) several noteworthy qualitative considerations surrounding these processes and their assessment.[20]

As the Canadian document notes, analysis 'begins with the presumption that a threat potential exists'. This work is done 'in order to develop an indicator list', which offers 'predictive utility' as to what to be on the lookout for or to be sensitive towards.[21] That indicator list also 'becomes a key mechanism by which we evaluate a threat potential.'[22]

Operating in a constantly looping feedback manner, the 'SEW process' is summarised in the Canadian document as consisting of and involving the range of activities 'from threat perception, to evaluation and monitoring, to assessment and warning.'[23]

---

[19] Europol (2013) see also Edwards (2013).
[20] Criminal Intelligence Service Canada (2007). Another valuable document worth citing here is: Quarmby (2003).
[21] See also, e.g., Svendsen (2012b).
[22] Criminal Intelligence Service Canada (2007). p. 6.
[23] Ibid., p. 8.

3.1.1 The Utility of 'Indicators'-Based Approaches

Providing some further detail as to how to best categorise what to be on the lookout for or what to be 'sensitive' towards, the document notes that: 'Indicators generally fall under one of two categories: primary indicators (or agency indicators) and secondary indicators (or structural indicators).'

Here, 'Primary indicators are those directly relating to activities (or transactions) of target individuals or groups…', while 'Secondary indicators… constitute the conditions that would either enable (make possible) or promote (make more likely) something to occur.'[24] These 'indicators' are referred to in the above referenced EUROPOL SOCTA document as 'facilitating factors'.[25]

Furthermore, the observable 'activities or transactions … may emit identifiable signatures … that can be associated to a particular group or actor…', and 'when transactions and signatures are tracked over time and interpreted in context', for instance, in relation to discernible patterns, 'they can provide critical insight into the capabilities and intentions of a group or individual.'[26] For example, providing emergent foreground-ranging 'signals' out of background-ranging 'noise'.[27]

Offering further insights regarding the 'monitoring' features and why they figure as an essential part of the overall process, the Canadian document remarks: 'regular monitoring and re-assessment are integral features of any strategic early warning system…'[28]

Starting the environmental scanning process from the most beneficial footing: 'The best indicator lists are those that address all the key threat components…', thereby offering 'a more targeted and efficient collection effort to evaluate what indications, if any, are present, and thereby to come to the best possible judgement on the threat level of a scenario.'[29] Moreover, 'The evidentiary support for indicators will be stronger for some than for others…', with a different colour-coded key both used and recommended as the most expeditious way 'to reflect the degree to which the evidence supports the finding that the indicator condition is present.'[30] Suggesting the strength of this 'traffic-light' (red-amber/orange-green)-characterised approach, 'Colour-coded scales are one means to convey meaning quickly and intuitively…'[31]

Underscoring the 'plurality' of interactions involved, the Canadian document valuably reminds us of the advantage of using different scenarios as a

---

[24]'Ibid., p. 11.
[25]See, e.g., EU Council (2012), p. 10.
[26]Criminal Intelligence Service Canada (2007), p. 12.
[27]Silver (2012); see also as discussed earlier in this chapter.
[28]Criminal Intelligence Service Canada (2007), p. 13.
[29]Ibid., p. 14.
[30]Ibid., p. 18.
[31]Ibid., pp. 20–21; see also further discussion of the 'traffic-light' system and its use, below.

methodological approach, with that approach 'provid[ing] intelligence consumers with the spectrum of alternative threat possibilities.'[32]

Highlighting some of the challenges involved in serious and organised crime contexts, the Canadian document notes: '…warning analysis for criminal intelligence must contend with a far greater number of variables and a conceivably limitless array of possible outcomes…' than, for example, when compared with traditional military warning processes. In that last context, distinct 'lines' might be somewhat clearer (evidence of military troop and equipment/materiel build-ups in particular geospatial areas, etc.).[33]

Bringing some greater overall clarity, 'crisis situations are often the culmination of a series of events and conditions, some of which will generate detectable signals or warning indications….'[34] Moreover, as the Canadian document goes on to note: 'strategic early warning is… concerned with the unknown or unexpected dangers over the horizon…'[35] These can be summarised as the 'green-coloured' issues (meaning that the events and developments being judged can be allowed to continue non- or less(er) addressed), rather than the 'red-/amber (orange)-coloured' or more immediate 'alert' issues (meaning that events and developments need to be stopped or responded to in some manner—e.g., including being changed, disrupted or frustrated in some form). More sustainable strategic rather than crisis management approaches are required in the less urgent contexts.

3.1.2 Limitations and Further Challenges

Several 'limitations' and 'challenges' with the overall SEW process are likewise appropriately noted. Indeed, the Canadian document ranks 'The most fundamental challenge' as being 'an epistemological one' (or the 'knowledge problem'), which in detail 'stems from the very nature of the warning mission: the future is inherently unknowable and yet we are trying to speak with some degree of confidence about events that have yet to occur.'[36]

Demonstrating that keeping an 'open-mind' is a useful characteristic to maintain throughout the whole SEW process, advanced and sophisticated, yet suitably focused, analytical processes must be maintained. Notably, this is 'so that a clear judgement'—that as closely as possible adheres to the thorough meeting of the 'end-user requirements criteria' of STARC (as already introduced above and discussed further below)—'can be presented to the intelligence consumer on the potential impact and likelihood of a given threat scenario.'[37]

---

[32]Criminal Intelligence Service Canada (2007), p. 19.
[33]Ibid., p. 7.
[34]Criminal Intelligence Ibid., p. 7.
[35]'Ibid., pp. 7–8.
[36]Ibid., p. 22.
[37]'Ibid., p. 22.

Several of the—even paradoxical—challenges confronted in the SEW enterprises are similarly highlighted, continuing to reflect their noteworthiness: 'we are, in a sense, left wandering in the dark, constantly looking for the "unknown unknown"—the potential threat that we have not even contemplated, let alone identified. This invariably makes strategic early warning for criminal intelligence more ad hoc in its practice, and its topic selection more analyst-driven....'[38] The 'human' factors/dimensions, such as human input to the overall process(es), clearly cannot be more denied, reminding us of the widespread 'art' and 'science' nature of intelligence and its analysis, including extension into more 'engineering' realms.[39]

All of the above factors should be kept in mind when engaging in SEW processes. This is, however, those above factors might be precisely configured in their exact calibrations as they run and progress over time. Greater systematic rather than more compartmentalised qualities are constantly sought.

Several challenges and limitations have already been raised, particularly building on the valuable findings from the Canadian model example. Going beyond those useful introductory insights, many other challenges and limitations can also be cited. For example, during the data-mining of contexts for the purposes of Strategic Early Warning (SEW)—including attempting to communicate conditions of 'context appreciation' and, as frequently used in military terminology, 'situational awareness' or the more advanced 'situational-understanding'—the question soon arises: what do we precisely want from our interaction with 'Big Data'? As Viktor Mayer-Schönberger and Kenneth Cukier, argue, when engaging in 'Big Data' work:

> The correlations may not tell us precisely why something is happening, but they alert us that it is happening... Big data is about what, not why. We don't always need to know the cause of a phenomenon; rather, we can let data speak for itself.[40]

Undertaking this 'Big Data' work, so configured, provides us with answers to the 'what' question. This is the key area to begin looking at, so we know what to focus on, and, subsequently, which entities to target further such as using, through adoption, 'target-centric' intelligence analysis approaches.[41] Notably, this is whereby, as its pioneer, Robert M. Clark, describes: 'the goal is to construct a shared picture of the target, from which all participants can extract the elements they need to do their jobs and to which all can contribute from their resources or knowledge, so as to create a more accurate target picture.'[42]

However, an argument can be made here for the position that answering the 'what' question is good enough at this early stage. This is because, with a close eye to next possible steps, it paves the way for enabling us to 'dig deeper' and to start

---

[38]Ibid., p. 23.
[39]Svendsen (2015a), p. 59.
[40]Mayer-Schönberger and Cukier (2013), p. 14.
[41]See, e.g., Clark (2012).
[42]Ibid., p. 13.

looking at other, overlapping or reinforcing critical interrogative questions and other context building-blocks during analysis and assessment—for example, who?, how?, why?, when?, where?, etc., so that we can build up a 'big(ger) picture' and form an understanding (that is, make sense) of what is actually going on. Again, this is across a wide range of time and space horizons, which is useful for improved situational awareness purposes and so forth.[43]

Further work heading in those directions is clearly beneficial to many different stakeholders, operationally and strategically alike.[44] Ultimately, summarising the challenges confronted in these 'early warning' and similar contexts, Florence Gaub of the EU Institute for Security Studies notes: 'early warning and preparedness face a number of challenges which derive partly from the essence of being human (i.e. the avoidance of cognitive dissonance), are partly the result of our strategic environment (i.e. differing assessments of a crisis [(alternatively, one can also read "issue"/"problem"/"threat"/"hazard" and "risk" here)]), and are partly the outcome of a broader institutional landscape. Recognising these challenges is most certainly the first step to addressing them.'[45] Examined next are the details of models which are relevant to conducting environmental scanning for foresight relating to the future of crime.

3.2 Unpacking the Environmental Scanning Models

Recent literature on the subject of 'Organised Crime Typologies' places emphasis on different classifications of crime that have been developed within the field. Most classifications are concurrent with the typology developed by the United Nations Office on Drugs and Crime (UNODC) in 2002.[46] Largely, organised crime can be categorised into three typologies:

1. Physical structure—for instance, identifying the key actors in criminal networks.
2. Activity of organised crime groups, hereunder profit-oriented activities. The business model approach exemplified below highlights these activities.
3. Conditions that have influence on organised crime activity. For example, social, historical and cultural. Facilitating factors found in the PESTLE model highlight these indicators.

Using this typology, and the herein associated methodologies combined with a conventionally recognised SWOT (strengths, weaknesses, opportunities and threats) analysis, collectively appears to represent a possible new approach to

---

[43]See also Symon and Tarapore (2015), pp. 4–11; Meserole (2016).
[44]See, e.g., as discussed in detail in Svendsen (2015a), pp. 58–73, and Svendsen (2015b), pp. 105–123.
[45]Gaub (2014), p. 82; see also insights and discussions in Gray (2014).
[46]See also insights from (UK HMG 2013).

environmental scanning. That approach has potential for being adopted and contributes additional information to the overall field.

3.2.1 The Micro-, Mezzo- and Macro-Levels of External Environment

The external environment or 'space' in which OC operates exists at several different levels. In this regard, there are multiple external environments. On the one hand, this can be seen at the micro-level, where the micro-environment relates directly to individual criminal activities and circumstances or to particular groups. On the other hand, the macro-environment consists of the widest possible framework of operation. This includes market spaces, related industries, or other broad networks found at the global or international level, where changes may directly or indirectly have an effect on OC groups' activities. Lastly, the mezzo-environment represents all that exists in between—that is, beyond the micro-environment, yet simultaneously less encompassing than the macro-environment. The mezzo-environment largely relates to processes of operation and to organisational specifics.[47]

It should be noted that propositional knowledge exists within the PESTLE scheme (see below) and is relevant at all levels—albeit typically on the macro-level. That propositional knowledge remains dynamic in nature and is highly contingent. It varies across time and space, and thus impacts our procedural knowledge. For example, changes may originate from the micro-, mezzo-, or macro-environments from a multitude of areas such as technological advancements, economic trends, or political developments, and do affect the other levels of information. For predictive policing, focus on the high-level environmental factors and maintaining a robust personal data protection threshold for privacy purposes is beneficial for a significant range of stakeholders.[48]

3.2.2 The PESTLE Model

Out in the 'real world', there are various types of knowledge and information we can obtain. In essence, managing—gathering, analysing, and interpreting—such a wide scope of information about our external environment is an arduous task, often, to at least some extent, represented by familiar 'intelligence cycle' processes. The PESTLE—in its detail covering political, economic, social, technological, legal, environmental factors—allows for the divergence of knowledge into different spheres of information that together overall compose a comprehensive outlook and understanding of our situated existence.

Since the PESTLE was originally developed for business purposes, it is basically a way to frame the primary business areas that should be evaluated. Traditionally,

---

[47]Klerks and Kop (2008), p. 20.
[48]See also insights in Svendsen (2015b), pp. 105–123.

the PESTLE schema asks questions and proposes themes that serve as guides for environmental analysis (Table 3).

Table 4 shows how indicators can be ordered according to the PESTLE domains. This is by also taking a macro-, mezzo-, micro-approach and moving from a macro-level strategic approach, which closes in on the tactical and operational dimensions and acknowledges the personal data protection issues that become more important the closer to the operational domain the analysis work gets. It is worth noting that the primary focus here is strategic, this is rather than being merely so operational/tactically orientated (Table 4).

3.2.3 SWOT Analysis

A SWOT analysis is a decision-making tool that is formulated by analysing, for instance, an organisation in comparison with its competitors. The analysis is structured by evaluating the strengths, weaknesses, opportunities and threats of or to an organisation. The purpose of evaluating the four elements is to identify the external and internal factors that are constructive or not, in order to achieve a specific objective. Hence, a SWOT analysis enables an organisation to enact a strategy that considers both the internal and external factors of an organisation. The internal factors are defined as the strengths and the weaknesses within the organisation, whereas the external factors are the opportunities and threats outside the organisation.

In order to examine the internal factors—strengths and weaknesses—it is useful to evaluate the types of advantages organisations have in comparison with their competitors, and also where there are areas of improvement within the organisation. External factors—opportunities and threats—contain macroeconomic issues such as socio-cultural change, changes in technology, or new government policies that would affect the organisation.

The fundamental outcomes of the SWOT analysis are converting and matching activities. Converting means to apply a strategy that converts threats or weaknesses into opportunities or strengths. Matching, on the other hand, enables the organisation to discover competitive advantage by strictly matching the strengths to opportunities. Hence, the organisation becomes better situated to achieve its objectives, and, furthermore, it allows for the organisation to avoid or, at the very least, minimise its weaknesses and perceived threats (vulnerabilities). Many common risk management aspects come to the fore.

Strengths, weaknesses, opportunities and threats (SWOT) analysis is again highly dependent on specific contexts, with assessments proceeding on a case-by-case basis following the identification of distinct 'targets' and then following 'target-centric' intelligence analysis approaches (as introduced earlier).[49]

---

[49]For example, see Danmarks Radio (DR) News (14 April 2014); DR News (10 April 2014); DR News (14 March 2014).

Table 3 PESTLE

Political	Economic	Social	Technological	Legal	Environmental
Major conflict and geopolitical developments	Who produces goods and services (means of production)?	Demographics	Technology available	What legal considerations must be taken?	
Terrorism	Distribution of goods and services	Human development (income, education, health)	Who uses technology?	What laws govern our behaviour?	
US leadership	Who are the consumers?	Cross-border migration	What impact does technology have?	Who creates and/or enforces the laws?	
China as global power	How are economic transactions defined?	Urbanisation (population density)	Access and competencies related to technology	The strength of legislation in combating serious and organised crime	
Towards multilateralism	Who holds capital?	Education		Loopholes in current legislation	
New global players		Trust in government		The ease with which new legislation can be passed	
Political stability		Corruption		The review process for current legislation	
Effectiveness of public administration		Voting trends: how is society organised?		International legislation and standards and their impact on national issues	Who is responsible for the environment?
		What are the characteristics of the population?			Are our practices sustainable?
		Where do people live?			Do environmental circumstances affect our behaviour?

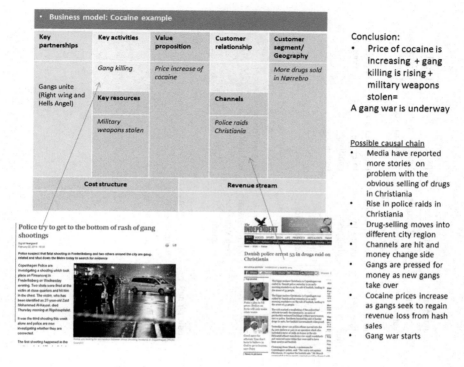

Fig. 1 OCG activity as business model

3.2.4 Business Model Approach

According to Albanese (2011), organised crime groups (OCG) are like any other business and will behave according to rational business needs. Thus, they are governed by market dynamics, enterprise, and the pursuit of profit. Helpfully, they can then be analysed using conventional economic models.[50]

The business model approach towards analysing OCG fundamentally sees an OCG as *any other type of business* (see Fig. 1). This means that OCGs are governed by a business model which involves key partnerships and channels in a value chain that supports the aim of the business, and enables it to deliver value propositions to its clients. An OCG is organised in a particular way and maintains key resources in the form of competent individuals that enables the OC business to provide value for its clients with minimal interference from the law. The OC groups serve a particular segment of clients that have distinct characteristics and needs which are based on the factors that define the value proposition of the OCG (Table 5).

---

[50]Albanese (2011).

Table 4 Vertical PESTLE

		Political	Economical	Social	Tech.	Legal
Strategic	EU international level	Est. OCG groups in EU	EU economic indicators	No. of people living in ghettoes in EU	New technology on the horizon	International law
	Macro-data (country level)	No. of members in OCG in country	Consumer confidence index	No. of people living in ghettoes	Price and access to crime enabling technology	Legislation in country
	Regional/mezzo-level/community level	No. of members in OCG in region	Unemployment rate in given area	Ghetto cities/region		
Tactical/operational	Micro-level data	Data associated with spec. groups		Particular social housing area with problem	Access to technology	
Personal data protection						
	Personal data	Political sentiment	Credit rating, unemployed	Spec. person living in spec. ghetto		

Table 5 Business model crime approach

Business model: Cocaine				
Key partnerships	Key activities	Value proposition	Customer relationship	Customer segment/ Geography
• Corrupt officials • Gangs • Supply partners in Africa, Mexico, Colombia	• Smuggling • Money laundering	• Value for money • Safe (pure)	• Demand creation	Typically more high-level users that can pay (Bankers, consultants more exposed to blackmail)
	Key resources		*Channels*	
	• Competent people • Financial strength		• Safe distribution network	
Cost structure			*Revenue stream*	
• Cost driven-scale advantage • Variable cost			Product sale, "licensing" others to sell	

Behaviours of OCG can be monitored and interpreted by using business logic. Any OCG can be described as having relative strengths and weaknesses, as is defined by the business model approach. As is presented in Fig. 1,—the example of gangs and organised crime as manifested over recent years on the streets of Copenhagen, Denmark, can be used to effectively illustrate the relations between their dynamics and the types of business models intimately involved in the gangs' OC activities. The example is illustrative only.

3.2.5 Definition of Indicators and Facilitating Factors

According to the Merriam-Webster dictionary, an indicator is 'a sign that shows the condition or existence of something'. In this regard, an indicator mirrors the ancient Greek definition of a sign: *'aliquid stat pro aliquot'*—something stands for something, e.g.: a red light functions as an indicator not to pass the pedestrian crossing. The properties of an indicator can take different shapes, but often they are limited to information gathered through different sources that convey something (e.g. knowledge) about something else (e.g. trends), which is relevant to OC analysis. Accordingly, indicators are pieces of information that can be expressed in different ways, such as amongst each other; and they may have a numeric or a non-numeric value. Typically the term 'indicator' in this context is reserved for numeric information used for measuring change in another pertinent piece of information, often described as a trend.

Historical lessons and the concept of 'surprise' teaches us to acknowledge bias. A change to the *status quo* is often realised too late, and thus indicators used for early warning need to take this into account. Indicators can be compiled based on identification of historically pertinent pieces of information or be based on scenarios. The latter approach is particularly important for early warning as they consider the role of surprise. Hypothetically, for example: If OCG would enter a neighbouring country, what behaviours and conditions would we expect to see? (CISC 2007) On the basis of these assumptions, a list of indicators is compiled that may look different to the indicators obtained when monitoring OC under normal circumstances.

Including rather than thinking merely in terms of more 'static' indicators, we also need to think here in terms of more fluid and (even rapidly) morphing 'event and development dynamics'. As a recent, prominent RAND report focusing on 'Predictive Policing' and its associated methodologies and tools for facilitating that approach remarked in 2013:

> Methods for predicting the illegal activity of organized crime groups include using the activity and dynamics of the group to evaluate risks and techniques to measure criminal market opportunities. The most common models include using group activity to identify organized crime, determining the presence of organized crime through illicit markets, and using a risk-based assessment (a more holistic approach) to identify illicit behavior.[51]

Continuing, and defining some readily recognisable operational parameters for this type of 'framing of the problem' being examined, the RAND report noted: 'For the most part, to date, these methods have been qualitative rather than formal statistical models [(i.e. quantitative)] for calculating the risk of "organised crime."'[52]

In the EUROPOL SOCTA approach, a distinction is made between effect indicators and crime relevant factors (CRF) also termed 'facilitating factors' (see above). The definitions thus follow a cause and effect differentiation, which in a dynamic system is not as fixed as one would like, as feedback loops and dynamic properties can turn an effect of one crime into the cause of another.

According to the SOCTA methodology (see also above), effect indicators are used to measure the impact of OCG and crime areas in the EU and on the European economy more widely. These indicators are an important tool in estimating adverse impact and are fundamental to evaluating and prioritising which actions have to be taken to prevent OC. These indicators thus serve as a tool for monitoring the success of OC crime prevention and are a vital tool for informed policy- and strategy- to decision-making.

Also, according to the SOCTA methodology, crime relevant factors (CRF) or 'facilitating factors' (see above) are used to determine vulnerabilities in society that may or may not lead to an increase in crime or how crime operates. CRF thus provide input for an opportunity assessment tool, e.g. the SWOT analysis, which assesses the ways that OC may differentiate and grow in the future by taking advantage of those vulnerabilities in society.

The facilitating factors can, for example, be unemployment and are found under the PESTLE classification model. The facilitating factors can be factors that

---

[51]Perry et al. (2013), p. 97.
[52]Ibid.

analysts have found to be conducive to OC. Facilitating factors are not fixed, but vary as they are highly contextual. As an example, the increase in unemployment is in some areas indicative of OC, as unemployment can serve as a facilitating factor. However, in other areas, the indicator unemployment does not serve as an indicator of OC, and so it is not a facilitating factor.

There are several typologies of indicators dependent on their use. Indicators can be topic based, longitudinal or ontologically defined. Indicators can also be further subdivided within these categories. Examples of indicators based on topic could be indicators that are divided according to high-level topics like violent crime, property crimes or white collar crime. These thematic areas comprise the sentinel watch list and its associated indicators (as discussed earlier, above). Further insights into the different indicators and their properties are now provided in turn below:

Longitudinal Indicators

Composite indicators: Composite Indicators are also known as indexes. They are compiled indicators that historically have shown themselves to be pertinent in the area that is being studied. In the organised crime context, a central component identified in the RAND study—and also found in and from other studies, including some the RAND study itself draws on—is the 'entrepreneurial(-ism)' factor. Notably this includes the core observation that 'fluctuations in the [(illicit and licit)] market space explain the behaviour of criminal groups.'[53]

Leading indicators: Leading indicators are indicators that are indicative of future events or trends. An often-cited leading indicator for the world economy is the Baltic Dry Index. In criminology, unemployment could be a leading indicator of increases in crime.[54] Statistically based early warning systems (EWS) employ a specific set of leading indicators often using graphical analysis as first indication of the early warning qualities of a leading indicator.

Coincident indicators: Coincident indicators are indicators that coincide or move in step with the development of the subject field. Coincident indicators are an important tool in a monitoring process, especially in fields where the subject under study does not yield any indicators that are easy to retrieve, which are directly related. If less expensive or easier to come by, historically coincident indicators can be obtained, these can serve as a proxy indicator. In crime prevention, the number of arrests does not necessarily indicate more crime, but could indicate greater resources allocated to criminal justice efforts. As such, coincident indicators on crime escalation can serve as a monitor on the progress of the criminal justice efforts as opposed to the number of arrests themselves.

Lagging indicators: Lagging indicators are indicators that transform after the subject under study has changed. Lagging indicators do not serve as a tool for early warning, yet can be important indicators to validate conclusions about, e.g. the increases in crime rates or other changes. While lagging indicators cannot serve as

---

[53]Perry et al. (2013), p. 97.
[54]See, for instance, relating to the 'lack of opportunity', Ratcliffe (2016), p. 163.

predictors, they can yield a greater understanding of the entire process and field of study.

Procyclical indicators: Procyclical indicators are mostly used in economics as indicators that move in the same direction as the subject under study.

Countercyclical indicators: Countercyclical indicators are also used mostly in economics and move in the opposite direction as the subject under study. An example of a countercyclical indicator in the area of OC could be an indicator of feelings of safety in a community related to the overall crime rate in the same community—feelings of safety decrease when crime increases.

Qualitative and Quantitative Indicators

Quantitative indicators: Quantitative indicators are statistically based numeric trends serving as quantitative measures of crime, for example, the number of drug-related criminal incidents in a country.

Qualitative indicators: Qualitative indicators are indicators that are derived from using a qualitative method in the study of crime. Qualitative indicators can be derived from informants (e.g. 'CHIS'—Covert Human Information Source(s)—as used in the UK), observation, investigations or interrogation. These indicators are often more subtle and singular and related to specific events. However, they can be used to describe overall trends in the crime community, which no valid data can account for—e.g.: information gathered from this source during an investigation can indicate that a change of MO (*modus operandi*) is taking place. The potential also exists for making many comparisons.

3.2.6 Process for Identification and Prioritisation of Facilitating Factors

(a) **Identification**

Identification of facilitation factors will typically be a (highly qualitative, even subjective) job done by law enforcement practitioners with intimate knowledge of OCG. Frequently by virtue of their experience, even following 'gut-instinct' or intuition, those practitioners know and understand what drives different OCG and can provide best positioned insights into their modus operandi. However, the more quantitative and objective tool of data-mining can also lend some valuable insight that may support law enforcement in their analysis of facilitating factors.[55]

A specific form of data-mining known as clustering is particularly useful. Clustering algorithms form a class of data-mining approaches that can be used to explore commonalities across crimes. This method can be used to find modus operandi for an OCG: e.g. a number of fishing boats are found to be smuggling cocaine. This may be a sign of a new MO for an OCG, and if a fisherman is found

---

[55]Ratcliffe (2016), p. 15.

to be smuggling in the future, this could be a sign that he is working together with that particular OCG.

If the commonalities found in the data-mining predate the crime, they may be leading indicators of the crime and thus crime facilitators. In practice, this approach emerges as being most strong when comparing and contrasting particular datasets to reveal pertinent pieces of information. The problem with this approach is that the results depend heavily on the input data and its quality, veracity, etc. The safeguard of traceability is required to be able to adequately oversee the anticipated vulnerabilities with this methodology.

It is fairly straightforward to find out if robberies tend to happen on sunny days by comparing and contrasting dates of robberies with dates of sunny days. It is a rather more different exercise to compare robbery data with some more obscure and abstract ranging or even more inaccessible piece of information such as whether parents were alcoholics. This is because the latter piece of information, even though it might be pertinent, would not (necessarily) be registered or might be recorded elsewhere in a different—maybe not even compatible or interoperable—format.[56]

(b) Process: prioritisation of weak signals and facilitating factors

The concept of weak signals entails that the weak signal is an antecedent to the event (see as introduced above). However, it predesignates factors that are not necessarily understood in a conditional or causal fashion. In Ansoff's understanding, the weak signal is weak because of the high level of uncertainty or the obscurity involved. Thus prioritising the weak signal becomes vital for operation. This prioritisation is done by looking at the weakness of the signal and by the consequence of what it refers to—the event or crime in question. The weakness is in part assessed by using the STARC framework approach.[57]

The signal should optimally adhere to a number of maxims that govern communication. If these maxims are not sufficiently adhered to, then there is the risk that vital pieces of information are disregarded. The weak signal will, by definition, not adhere to some of these underlying maxims found in the STARC approach (see Table 6). Two other aspects of the weakness of the weak signal is the predictive capacity of the indicators used and the credibility of the source.

Ultimately, the '3Rs' requirements need to be effectively met. Namely, this is *'getting the right information, to the right person/people, at the right time'*. For the conditional weak signal, which functions as a facilitating factor, a risk assessment needs to be made. Three factors are important: actors, conditions, events.

---

[56]See for these sorts of issues and challenges, e.g., I.R. Porche, III (2014), insights as discussed throughout Clark (2012).

[57]See these criteria in Hall and Citrenbaum, *Intelligence Collection* (USA: Praeger Security International—PSI 2012), p. 81.

Table 6 STARC

Specific(ity)	Information should be specific enough for a response to be made. Addresses the 'what?' question
Timely (timeliness)	The concept of early warning equally implies that, on the other hand or on the other side of the coin, there is such a thing as warning that comes too late and thus is not timely. A warning must come in due time (at the 'right' time) for an operation to be able to respond to the risk which the weak signal forbears
Accuracy (reliability)	Information needs to be accurate, moving us also into the reliability of sources and data domains and associated considerations
Relevance	Any piece of information needs to be relevant (e.g. what a source is reporting). Addresses the 'so what?'/'what does it mean?' questions
Clarity	Ambiguity needs to be cleared up as far as possible, yet still enabling complexities to be adequately captured and communicated

Table 7 Risk level for OCG primary indicators

Risk level for OCG indicators		
	Low	Co-operation with other groups, expertise, external violence, countermeasures against law enforcement
	Medium	Adaptable and flexible, level of resources, the use of legal/legitimate business services as a front for criminal activities, active in multiple crime areas
	High	An international dimension to their activities, the use of corruption

Risk level for SOC area indicators		
	Low	Limited resource availability, social tolerance, linked crime areas (e.g. local)
	Medium	Innovation, number of groups active and evolution of the crime area (e.g. OCG linked regionally)
	High	International dimension (e.g. OCGs linked globally) and high profits

Source Europol SOCTA

Primary Indicators (Table 7).
Secondary Indicators (Table 8).

By multiplying the impact (on a range 1–3; low to high) of an event with the probability (including conditional probability) a risk assessment can be made.

Table 8 Risk level for OCG secondary indicators

Events		
	Low	Slow; far; peripheral; small-size; it is an… issue; it is a… risk
	Medium	It is a… problem; it is a… hazard
	High	Fast; near; close; centre/core; large-size; it is a… crisis; it is a… threat
Conditions		
	Low	Early warning; small volume; weak impact/influence; long-range (1: 0–33%); pursue 'intelligence methodology' (wait and watch)
	Medium	Warning; mid-range (2: 34–66%); pursue hybrid intelligence and security method of 'assess, evaluate, monitor, and—if required—disrupt'; 'risk threshold' zone (e.g. where 'red-lines' are drawn)
	High	Alert; large volume; strong impact/influence; Short-range (3: 67–100%); pursue 'security methodology' (see and strike)

Source The authors

4 Drawing Dimensions Together

In order to utilise the different strengths that each environmental scanning model possesses for foresight and for gaining insights into the future of crime, the output of the model needs to be fused with the output of the other models. This 'source-triangulation'-related work is done in order for the models to supplement each other and to gain a more refined threat assessment. Overall understanding to knowledge is enhanced through the different methods contributing with different types of information to the overall analysis.

4.1 Combining the SWOT and the Business Model Approach

In summary, the business model approach guides analysts to conclusions regarding the relative strengths between different actors involved in organised crime activities. On the one hand, the Hells Angels gang have the key resources and the financial strength to push the Loyal to Familia gang into an expensive gang war. However, the Hells Angels lack the (necessary) support from the ethnic community in a very important area of Copenhagen.

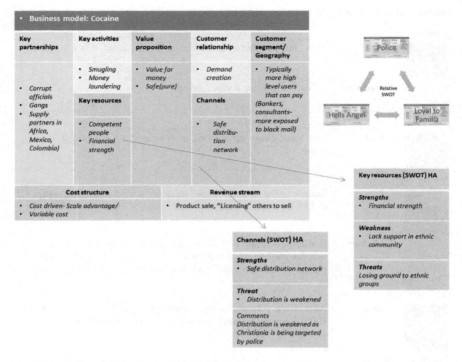

Fig. 2 Analysis of OCG activity with SWOT and business model

As their distribution network continues to be targeted by police, and as the OCG continue to lose market share in significant districts, their business suffers losses. They, therefore, face future challenges. The rational choice is to consider engaging in gang warfare by conducting a cost-benefit analysis, using conventional methods. The Hells Angels have the financial strength to go to war and can leverage their experience and strong network to their advantage. Loyal to Familia is a new OCG and a competitor to the Hells Angels. It holds a strong position in an important region in Copenhagen that has traditionally been controlled by the Hells Angels, but they lack experience and key partnerships. There is strong motivation for Hells Angels to wage war on Loyal to Familia—for example, to protect their market—and an analysis of these signals support the notion that this should be monitored (Fig. 2).

4.2 Combining the PESTLE and the Business Model Approach

In the example cited above, the business model approach indicates that a motive exists for Hells Angels to go to war with Loyal to Familia, so a gang war is possibly

		Political	Economical	Social	Tech.	Legal
Strategic	EU International level	Est. OCG groups in EU	EU economic indicators	No. of people living in ghettoes in EU	New technology on the horizon	International law
	Macrodata (Country level)	No. of members in OCG in country	Consumer confidence index	No. of people living in ghettoes	Price and access to crime enabling technology	Legislation in country
	Regional/ Mezzolevel/ community level	No. of members in OCG in region	Unemployment rate in given area g	Ghetto cities/region		
Tactical/ operational	Microlevel data	Data associated with spec. groups		Particular social housing area with problem	Access to technology	
			Personal data protection			
	Personal data	Political sentiment	Credit rating, unemployed	Spec. Person living in spec. ghetto		

- **Business model: Cocaine indicators**

Key partnerships	Key activities	Value proposition	Customer relationship	Customer segment/ Geography
• Gangs unite	• Gang killing	• Price increase of cocaine		More drug sold in Nørrebro
	Key resources		**Channels**	
	• Military weapons stolen		• Police raids Christiani	

Fig. 3 Combining the PESTLE and the business model approach

beginning (or, at least, can be anticipated). Consequently costs will increase. The PESTLE contributed to this analysis emphasised that consumer prices are falling, resulting in even further falling revenues. If war is underway, the gangs need to diversify—they recruit.

The recruitment potential is assessed by the PESTLE model, which provides information relating to the degree of ghettoisation and unemployment. In unison, these last factors provide the information that supports the overall early warning being issued as described in the models described earlier in this chapter (Fig. 3).

Ultimately, pointing to their more 'strategic' utility—rather than so much to merely their 'tactical/operational' deployability—as the RAND report on *Predictive Policing* observes: 'In general, the methods [(of 'Environmental scanning'; 'Nominal group and Delphi process'; and 'Scenario writing'—described in detail in the cited RAND report over pages 97–98, and including extending to the other

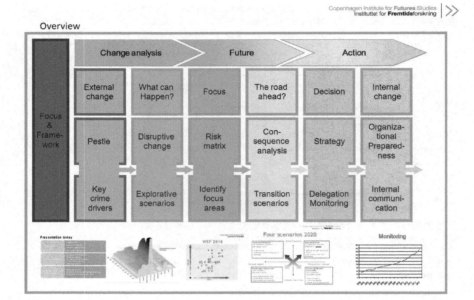

Fig. 4 Future scenario process overview

methodologies described and referenced throughout this document)] are designed *more to identify opportunities for crime than to monitor the nature of the group itself.*'[58] They should not become over-extended or stretched too far in their use. Adequate, balanced perspective needs maintenance.

5 Conclusions

From the discussion undergone during the course of the journey taken by this chapter, many overall conclusions can now be drawn. End-users in the expansive and diverse form of intelligence, crime and security analysts to higher-level commanders, customers, clients and consumers to other policy- to decision-makers can learn much in terms of at least starting or beginning 'tips' and 'leads' from harnessing the work discussed throughout this chapter. However, this is only part of the overall 'story'. Much is now ready for further investigative follow-up, such as for more robust—even prosecution-viable—evidence acquisition purposes, launching future intelligence-led operations, and so forth. Effective launch-pads can

---

[58]Perry et al. (2013).

clearly be provided. After all, as Gaub has observed: "'Pouring the data into a narrative" is the main task of risk analysis.'[59]

Finally, the rationale for engaging in and expending energy on examining these 'warning enterprises' at their most broad is also starkly obvious for foresight and providing insights into the future of crime. As Australian Professor Alan Dupont and US Professor William Reckmeyer have argued:

> Foresight buys precious response time and, in matters of national security, prescience provides a vital strategic edge over potential enemies, as well as an enhanced capacity for leadership in the face of common threats [(including ranging comprehensively across full-spectrum issues-to-problems-to-hazards-to-risks encountered and experienced)].[60]

Many different stakeholders to end-users rightly remain interested in these trends, and indeed in the—frequently extended—advanced and sophisticated processes surrounding their evaluation.[61] Further productive and progressive endeavours are open to be taken in this area of work into the future (Fig. 4).

References

Albanese, J. (2011). *Organized crime in our times* (6th ed.). Burlington, MA: Anderson Publishing.
Andrew, C. (2004, June 1). Intelligence analysis needs to look backwards before looking forward. *History & Policy*.
Ansoff, H. I. (1975). Managing strategic surprise by response to weak signals. *California Management Review, 18*(2), 21–33.
Clark, R. M. (2012). *Intelligence analysis: A target-centric approach*. Washington DC: CQ.
Criminal Intelligence Service Canada (CISC). (2007). *Strategic early warning for criminal intelligence: Theoretical framework and sentinel methodology*. Ottawa, Canada: CISC, Central Bureau. http://publications.gc.ca/collections/collection_2013/sp-ps/PS64-107-2007-eng.pdf. Accessed June 23, 2016.
Danmarks Radio (DR) News. (14 April 2014). *Biker gangs involved in economic crime*
DR News. (10 April 2014). *Københavnske politikere er klar til dialog med rockere*
DR News. (14 March 2014). *Police arrest 80 in drug raid*
Dupont, A., & Reckmeyer, W. J. (2012). Australia's national security priorities: Addressing strategic risk in a globalised world. *Australian Journal of International Affairs, 66*(1), 46.
Dixon, M. (2013). *Textbook on international law* (7 ed., p. 52). Oxford: Oxford University Press.
Dixon, P. (2016, June 7). How to predict the future? Look to the past. *Wired. Organised crime, EU OC definition—ENFOPOL 35 REV 2*—20131211115812170-0-eu.pdf via www.epoolice. eu. Accessed November 2013.
Edwards, C. (2013). *Is Britain the organised crime capital of Europe?*, RUSI Comment/Analysis (19 March).
EU Council. (2012). *Serious and organised crime threat assessment (SOCTA)—methodology*. 12159/12 REV 2 (p. 10). Brussels, July 4, 2012.

---

[59]Gaub (2014), p. 81.
[60]Dupont and Reckmeyer (2012), p. 46.
[61]See, e.g., inter alia, Grabo (2004); Habegger (2009).

European Commission Directorate General for Research and Innovation Socio-Economic Sciences and Humanities. (2014). *European Union Research in Foresight.* https://ec.europa.eu/research/social-sciences/pdf/project_synopses/research_in_foresight.pdf. Accessed June 23, 2016.
Europol. (2013). *SOCTA.* https://www.europol.europa.eu/content/eu-serious-and-organised-crime-threat-assessment-socta. Accessed June 23, 2016.
Gaub, F. (2014). Enhancing early warning and preparedness. In P. Pawlak & A. Ricci (Ed.), Crisis rooms towards a global network? (p. 82). The European Union Institute for Security Studies
Grabo, C. M. (2004). *Anticipating surprise: Analysis for strategic warning.* University Press of America
Gray, C. S. (2014). *Defense planning for national security: Navigation aids for the mystery tour.* Carlisle, PA: United States Army War College Press.
Habegger, B. (2009). *Horizon scanning in government: Concept, country experiences, and models for Switzerland.* Zürich: Center for Security Studies, ETHZ.
Hall, W. M., & Citrenbaum, G. (2012). *Intelligence collection.* Praeger Security International.
Heuer, Jr., R. J., & Pherson, R. H. (2014). *Structured analytic techniques for intelligence analysis* (2 ed.). Washington, DC: CQ Press.
Jojarth, C. (2009). *Crime, war, and global trafficking: designing international cooperation.* Cambridge University Press.
Klerks, P., & Kop, N. (2008). *Societal trends and crime-relevant factors: An overview for the Dutch national threat assessment on organized crime 2008–2012* (p. 20). Apeldoorn, NL: Politieacademie.
Lockwood, J. S. (2013). *The lockwood analytical method for prediction (LAMP): A method for predictive intelligence analysis.* London: Bloomsbury Intelligence Studies.
Martinet, A. C. (2010). Strategic planning, strategic management, strategic foresight: The seminal work of H. Igor Ansoff. *Technological Forecasting and Social Change, 77*(9), 1485–1487.
Mayer-Schönberger, V., & Cukier, K. (2013). *Big data: A revolution that will transform how we live, work and think.* London: John Murray.
Meserole, C. (2016, June 5). National security in a data age. *Lawfare.*
Perry, W., McInnis, L. B., Price, C. C., Smith, S. C., & Hollywood, J. S. (2013). *Predictive policing Washington DC.* RAND
Porche, I. R. III, World Report—US News. (1 May 2014). *Why "big data" can't find the missing Malaysian plane: The ongoing search for Flight 370 exposes the world's serious information gaps.*
Quarmby, N. (2003). Futures work in strategic criminal intelligence. *Conference paper.* Canberra (Australia).   http://www.aic.gov.au/media_library/conferences/evaluation/quarmby.pdf . Accessed June 23, 2016.
Ratcliffe, J. (2016). *Intelligence-led policing.* London: Routledge (2 ed.).
Schoemaker, P. J. H., Day, G. S., & Snyder, S. A. (2013). Integrating organizational networks, weak signals, strategic radars and scenario planning. *Technological Forecasting and Social Change, 80*(2013), 815–824.
Silver, N. (2012). *The signal and the noise.* London: Allen Lane.
Svendsen, A. D. M. (2012a). *Understanding the globalization of intelligence.* Basingstoke: Palgrave/Macmillan.
Svendsen, A. D. M. (2012b). *The professionalization of intelligence cooperation: Fashioning method out of Mayhem.* Basingstoke: Palgrave/Macmillan.
Svendsen, A. D. M. (2015a). Advancing "defence-in-depth": Intelligence and systems dynamics. *Defense & Security Analysis, 31,* 1.
Svendsen, A. D. M. (2015b). Contemporary intelligence innovation in practice: Enhancing "macro" to "micro" systems thinking via "system of systems" dynamics. *Defence Studies, 15,* 2.
Svendsen, A. D. M. (2017). *Intelligence Engineering: Operating Beyond the Conventional* (New York: Rowman & Littlefield/SPIES - Security and Professional Intelligence Education Series).

Symon, P. B., & Tarapore, A. (4th Quarter 2015). Defense intelligence analysis in the age of big data. *Joint Forces Quarterly—JFQ, 79,* 4–11.
UK HMG. (2013). *Serious and organised crime strategy.* London. https://www.gov.uk/government/uploads/system/uploads/attachment_data/file/248645/Serious_and_Organised_Crime_Strategy.pdf Accessed June 23, 2016.

Big Data—Fighting Organized Crime Threats While Preserving Privacy

Anne Gerdes

1 Introduction

Observations about big data are sometimes framed overtly optimistic, for instance, assuming that: "We can throw the numbers into the biggest computing clusters the world has ever seen and let statistical algorithms find patterns where science cannot" (Anderson 2008). Other times, critical voices are raised, as in Jonas and Harper, who pass sentence on big data for terrorism fighting by concluding that:

> Predictive data mining is not a sharp enough sword, and it will never replace traditional investigation and intelligence, because it cannot predict precisely enough who will be the next bad guy.

(Jonas and Harper 2006: 10)

Therefore, an initial clarification of the promises and perils of big data is in place (Sect. 2), before moving on to a discussion of how the increasing use of big data analytics for crime fighting purposes challenges privacy and societal trust. Here, the ePOOLICE system is highlighted (Sect. 2.1) as an example of a sense-making system, which is founded on a privacy-preserving design framework (for a detailed description of the ePOOLICE system, see chapter three in this volume: "The ePOOLICE Project"). As such, the system functions as a case, which runs through

---

The chapter is an elaborated version of a conference short paper: Gerdes. A (2015) *EPOOLICE Security Technology—Fighting Organized Crime Whilst Balancing Privacy and National Security*, which was presented at the 10th International Conference on Cyber Warfare and Security, March 24–25, 2015, South Africa, Krüger National Park. I am grateful for valuable comments from the conference audience.

A. Gerdes (✉)
Department of Design and Communication,
University of Southern Denmark, Kolding, Denmark
e-mail: gerdes@sdu.dk

© Springer International Publishing AG 2017
H.L. Larsen et al. (eds.), *Using Open Data to Detect Organized Crime Threats*, DOI 10.1007/978-3-319-52703-1_5

the chapter and frames the discussion of how ethical and societal values may be challenged. Consequently, the growing use of big data technology-led policing initiatives increasingly puts privacy under pressure and calls for a discussion of national security and citizens' right to privacy (Sect. 3). On the backdrop of these observations, it is argued that core issues should not be addressed as a strict dichotomy of realms, formulated in a clash between citizens' right to privacy as opposed to national security; rather we have to strike a balance between a national and an individual dimension of security. In continuation thereof, security can be defined as nonattendance of danger at a state level, as well as at a societal level with reference to the citizens forming the society (Raguse et al. 2008). Moreover (as discussed in Sects. 4 and 4.1), it is emphasized that trust plays a vital role in democratic societies in which citizens' and governmental authorities interact freely with one another in an open manner. However, such trustful relations are hard to build and constitute important goods in open societies, yet at the same, they may easily come under pressure if we do not carefully consider ways to balance technology-based crime fighting initiatives with citizens' right to privacy.

2 Promises and Perils of Big Data

In a crime fighting context, big data, quite often in the shape of open source data, may provide data-driven sense-making, which makes it possible to reveal small meaningful patterns from big data sets by discovering correlations between diverse crime indicators and enablers. As such, the ability to see patterns in data, and thereby reveal previously unknown links between enablers, holds promises that valuable insights can be distilled from big data and subsequently inform decision-making and policing initiatives, such as predictive policing (Perry et al. 2013), intelligence-led policing (Ratcliffe 2011), as well as strategic long-term crime forecasting. But, before delving into the role of big data for crime fighting purposes, it is important to provide an overall clarification of this omnipresent phenomenon, which is sometimes referred to as mere "hype" or the latest buzzword. Nevertheless, the notion of big data was originally framed by the so-called three V's of big data: high volume, high velocity, and high variety (Laney 2001) indicating scale, including the ability to carry out real-time analysis on big data streams, speed, and the mere fact that we are now able to handle data independently of form. Later, in the article *Analytics: The Real World of Big Data. How innovative organizations are extracting value from uncertain data* (Schroeck et al. 2012), IBM added yet another V, *veracity*, to underscore the priority of trustworthiness in relation to big data analytics. As an example thereof, reservoirs of big data often depend on open online data sources. These kinds of data sets are not sources of natural given objective facts but represent human-made products, and it is important to bear in mind that such data may be error-laden and uncertain. Also, *Value* and *Viability* have been added (Biehn 2013) to emphasize the role of data as a strategic asset. First mentioned seeks to establish how data add value to the business by, for instance, setting out to distinguish between core business data as opposed to data, which can be sold, shared, or data that the business needs to buy elsewhere. Likewise, *Viability* stresses the importance of settling what the organization seeks to achieve by using data, and whether it has access to the methods needed to gain new insights from its data pools.

In essence, big data is not primarily about data availability, but about capability, i.e., having available techniques and skills, which may reveal knowledge discovery opportunities hidden in big (often unstructured) data sets. By the same token, Hilbert emphasizes the importance of *analytics* in relation to big data:

> The full name of Big Data is Big Data Analytics. The notion of Big Data goes far beyond the increasing quantity and quality of data, and focuses on analysis for intelligent decision-making. Independent from the specific peta-, exa- or zettabytes scale, the key feature of the paradigmatic change is that analytic treatment of data is systematically placed at the forefront of intelligent decision-making.

(Hilbert 2016: 139)

Consequently, big data, by itself, is worthless without analytics facilitated by a range of different computational techniques and tools, such as data mining, data fusion, and data integration techniques. Hence, data fusion uses reduction techniques to enhance data mining by merging structured as well as unstructured data sets from multiple sources into a homogeneous representation, which makes such data sets suitable for subsequent processing by data mining tools. Likewise, in emphasizing the importance of retaining data, integration techniques assimilate large data sets to obtain a near-complete picture of a given environment by keeping the larger set of data (e.g., integration of costumer databases with databases with household information) (PCAST report 2014: 25). On this background, data mining algorithms may both enlarge the amount of data accessible for usage and promote sense-making by discovering patterns across large data sets. In addition, data mining is naturally rooted in computer science as well as in theories and methods of applied mathematics, statistics, and artificial intelligence, in particular, machine learning. Here, deep learning algorithms are developed as self-learning algorithms that, to a certain extent, can be viewed as capable of adapting to and grasping a given environment. Accordingly, such algorithms gradually build up domain knowledge by being trained on an initial data set with selected indicators, which reinforces preferred pattern matching strategies.

Data mining techniques and algorithms cast the scene and reveal certain meaningful data patterns, on the cost of others, which are sifted out (Kitchin 2014: 5). However, although data mining may disclose patterns and present relationships it "usually cannot tell the user the value or significance of these patterns." (PCAST 2014: 25). Thus, data do not speak by itself in isolation. Moreover, limitations in data material or use of biased algorithms may negatively influence an analysis. For that reason, social media big data-driven research cannot avoid reflecting on methodological issues, such as, for instance, whether Twitter data are representative —not all people tweet, some have several accounts, and some are "bots"—or how scarcity of historical data possible biases results (Boyd and Crawford 2012: 669).

Additionally, the use of advanced algorithms may impede validation of results, because the involved steps of interpretation and analysis are automated and mediated behind a veil of algorithms. Hence, the promises of having access to almost all data with right to a given topic or behaviour (Mayer-Schönberger and Cukier 2013: 12) may falsely convince us that there is no need for scrutinizing data through the lens of theory or applying a systematic approach to data collection. Nevertheless, promises of big data knowledge opportunities are epistemologically constrained, and we ought to "consider how the tools participate in shaping the world with us as we use them" (Boyd and Crawford 2012: 675).

Moreover, the resultant outcomes of big data analyses display correlations and do not reflect causal relations. As such, it takes domain expertise and situational awareness to make use of this kind of imperfect information and figure out which patterns are truly meaningful and add value to the task at hand. In a similar vein, since small meaningful patterns are not significant, but only indicative, there is a risk that big data sense-making systems may potentially erroneously frame ordinary citizens' behaviours in harmful ways. As an example thereof, machine learning may be applied to train a data mining algorithm to match behavioural features for criminal profiling by matching features present in large data sets. Nevertheless, without humans in the loop and context knowledge, the algorithm will likely identify false positives, persons, who match the given characteristics, but who still are not criminals. Hence, contemporary technology-led policing, based on big data analytics takes departure in information derived from statistical inductive processes and models to uncover previously unknown patterns, as opposed to more classical investigation by deduction. However, capability to combine more information from more diverse sources may possibly enhance knowledge opportunities by highlighting correlations between data, but this fact does not, by itself, lead to actionable information—"a statistical relationship between one factor and a greater crime risk does not necessarily mean that the factor 'causes' the crime." (Perry et al. 2013: 17). This observation is exaggerated by Jonas and Harper (2006), who refer to predictive data mining for fighting terrorism as: "unfocused, false-positive laden," and "a waste of national resources." (Jonas and Harper 2006: 9).

To sum up, the problem with big data is not primarily related to data processing or aggregation capacities, neither to the size of big data sets, or the manifold variations they come in, rather it has to do with the epistemological challenge of seeing small patterns in big data, which add value in a given context. According to Floridi, in quoting Platon, the big data game will be won by those "who know to ask and answer questions (Plato, *Cratylus*, 390c)":

> We need more and better techniques and technologies to see the small data patterns, but we need more and better epistemology to sift the valuable ones.

(Floridi 2012: 437)

The role of human domain expertise is a critical factor for successful big data-driven crime fighting initiatives. As an example, at the outset of the ePOOLICE project, law enforcement end users were actively involved in the semantic framing of context-sensitive crime indicators in exploiting fully the concept of crime hubs, crime indicators, and crime enablers (Brewster et al. 2014). Moreover, the use context of the system is situated at the strategic level, implying that the system does not support the operational level at all, but serves a pure preventive purpose in scaffolding sense-making activities carried out by law enforcement agents and analysts engaged in countering threats and acting proactively in dealing with upcoming trends in organized crime. In that respect, the system and the human analyst are thought to work in tandem: The system presents weak signals, but the human analyst decides whether the given pattern should possible be sorted out as representing important discontinuities in the environment, and hence interpreted as early signs of an emerging organized crime threat. As such, data are not viewed in isolation and cannot escape qualification by human analysts, whose interpretation of data rests on skills grounded in situational awareness based on domain expertise.

In the following section, the ePOOLICE case is presented on the backdrop of the crime fighting landscape, and it is emphasized that although the development of the system has been guided by a privacy by design philosophy, it is nevertheless important to tune in on a broader discussion of the ethical issues and societal impact of such systems (Sects. 3 and 4).

2.1 Fighting Organized Crime—Insights from ePOOLICE

On a regular basis, Europol has (since 2010) provided a so-called SOCTA report—European Serious and Organized Crime Threat Assessment, which presents recommendations for priorities in European crime fighting. Hence, for the period 2013–2017, the below-mentioned priorities in fighting organized crime have been highlighted:

> The SOCTA 2013 [EU Serious and Organized Crime Threat Assessment] identifies a number of key priorities, which, in Europol's view, require the greatest concerted action by EU Member States and other actors to ensure the most effective impact on the general threat. Facilitated illegal immigration, trafficking in human beings, synthetic drugs and poly-drug trafficking, Missing Trader Intra-Community (MTIC) fraud, the production and distribution of counterfeited goods, cybercrime and money laundering are the particular crime areas listed in this category.
>
> (SOCTA 2013, foreword by Director of Europol, Rob Wainwright)

Likewise, once a year, Europol's European Cybercrime Centre (EC3) presents the "Internet Organized Crime Threat Assessment (IOCTA)." The report emphasizes three subpriority areas for fighting cybercrime in 2016, namely child sexual exploitation, cyberattacks, and payment fraud (IOCTA 2015: 6), and it is in general noticed that:

> Cybercrime is becoming more aggressive and confrontational. Various forms of extortion requiring little technical skills suggest changes in the profile of cybercrime offenders, and increase the psychological impact on victims.
>
> (IOCTA 2015: 10)

Unfortunately, cybercrime develops apace and has led to further consolidation of well-known threats, such as crime-as-a-service business models, growth in Dark Net markets, and the increasing use of bitcoins, outside Dark Nets, as a currency among criminals. Moreover, new trends are seen within the area of child sexual abuse with the evolvement of commercial live streaming (IOCTA 2015: 12).

The SOCTA and IOCTA reports draw the landscape of organized crime with the purpose of informing priority setting and operational actions for EU law enforcement. However, from a long-range strategic perspective, it is also important to be able to predict future crime scenarios lurking in the horizon. On this background, the ePOOLICE case represents a big data sense-making system (realized as a prototype demonstrating a proof of concept solution), which seeks to improve the effectiveness of law enforcement agencies (LEAs) by supporting proactive strategic decision-making processes. Consequently, by implying sense-making techniques, the overall aims of the system are to alert LEAs to potentially significant changes in organized crime *modi operandi* before they mature. Open source big data analytics and ontological knowledge representation in combination with LEA domain expertise might improve LEAs' capabilities of fighting organized crime by ensuring that LEAs are well armed, use proactive planning for countering threats, and are able to detect and deal with discontinuities or strategic new situations, i.e., discover "weak signals" (Ansoff 1975), which can be sorted out as important discontinuities in the environment and interpreted as early signs of an emerging organized crime menace.

In a security perspective, the system exemplifies a tool for strategic intelligence-led activities focusing on modus operandi, hotspot locations, emerging crime patterns and future (mega-) trends. For that reason, personal data are not relevant in the information context of ePOOLICE, since the system does not aim at identifying individuals. Moreover, a privacy-preserving design framework has driven the system development process in order to address issues of anonymization of potential personal and, or, sensitive data. Correspondingly, other privacy safeguards, regarding system functionalities, have been installed as well, such as, for instance, considerations to permit management of different user-levels of authorization. In that respect, logging of system access and processes ensure tracking of possible unauthorized use of data and further support control procedures by internal data controllers, as well as by independent authorities.

In seeking to balance data privacy and data utility, the system design is based on a privacy-preserving system development framework supported by legal research (Callanan et al. 2009; De Marco 2014) and in compliance with EU data protection law. Still, the general increase in technology-led policing initiatives calls for a broader discussion of the impact of security technology systems at a societal level, particularly with right to the balancing of important ethical values, and national or public security.

3 Balancing National Security and Citizens' Right to Privacy

The concept of security is not easily defined but can be conceived of as dealing with well-being at state level and protection of democratic institutions. Over time, the classical state-oriented notion of security has changed toward a more individual-centered approach, emphasizing the integrity and security of the individual and protection from threats (Liotta 2002). The rise of a human security framework as necessary for individual, as well as national and global, security is reflected in *the United Nations' Human Security Doctrine* (2003) and in *the Human security doctrine for Europe* (2004). Within this framework, security can be defined as nonattendance of danger at a state level, as well as at a societal level with reference to the citizens forming the society. Moreover, the EU Security Strategy (2004) emphasizes the need to act proactively in dealing with key security threats (Raguse et al. 2008: 16-ff.). As such, prevention is the core goal of EU security policy implying a proactive stance toward unclear risks and threats, which stem from terrorism, organized crime, environmental disasters and diseases. In order to act proactively and predict future outcomes, one has to be able to reveal patterns of information in complex dynamic changing environments, and for that reason, dataveillance has become a central component in modern security policy. From this viewpoint, democratic states are faced with the challenge of striking a balance between two sides of security, formulated as absence of organized crime threats and preservation of the freedom and integrity of citizens as important preconditions for a flourishing democracy.

Issues of security and surveillance of citizens are frequently framed as a dichotomous clash between citizens' right to privacy as opposed to national security, implying a zero sum game in which more security is necessarily followed by less privacy. This observation may give rise to civil society concerns regarding the protection of privacy rights in relation to the growing use of technologies. But given that security is also an intrinsic value for human well-being at a fundamental level, we might move beyond the dichotomy between citizens' right to privacy and national security and instead conceptualize security in terms of interacting and mutually dependent dimensions of security, i.e., as individual security and public and/or national security (Raguse et al. 2008: 16-ff.).

To elaborate on this from a legal perspective (De Marco 2014; De Marco in Callanan et al. 2009: 189-ff.), any limitation to fundamental rights of privacy and personal data protection has to respect some basic principles in order to be legitimate and ensure that privacy is not violated. Consequently, limitations have to rest on a legal basis and must be formulated with such a degree of precision that it enables citizens to understand how their navigation and conduct in society are affected by the given limitations. Moreover, a restriction must pursue a legitimate aim, i.e., be in accordance with listed legitimate aims, formulated within each article of rights in *the European Convention on the Protection of Human Right and Fundamental Freedoms* (ECHR) as aims that justify interference. Furthermore, any limitation must correspond to a real need of society and must be seen as an efficient instrument (for instance, in relation to crime reduction and security). Finally, the principle of proportionality seeks to guarantee that the limitation is balanced to the aim pursued. In order to minimize the infringement of privacy rights and to assess the proportionality of a restriction, important issues to settle are whether the overall effect of the constraint is reasonable and whether it is the least intrusive mean available (De Marco 2014: 11-ff.). Here, to ensure that privacy is not violated, one must specifically see whether the requirements of necessity and proportionality of a potential privacy restriction are satisfied and whether the given system is based on an appropriate legal basis. Given these circumstances, it is important to balance both privacy and security by introducing proactive privacy-enhancing design principles throughout all stages of the system development process—for instance, by carrying out privacy impact assessments specifically targeting the challenges inherent to the development of the given system (see, for instance, De Marco 2014; Gerdes 2014). Similarly, in order to support privacy compliant use of crime fighting systems, it is pivotal to emphasize the need for ongoing privacy impact assessments and privacy educational initiatives for LEAs once such systems are implemented in an organizational setting.

3.1 Privacy

Privacy is usually regarded as an instrumental value, but of a special kind since it is also a necessary condition for the fulfillment of important human goals, i.e., intrinsic goods, such as autonomy (Johnson 1994; Benn 1971), friendships (Fried 1968), and the formation of social relations (Rachels 1975). Consequently, Benn's argument below encapsulates the importance of privacy for autonomous decision-making and for the formation of identity:

> The management of so complex a set of relations and the self-assessments and revisions implicit in it, would be quite impossible if one could not insulate one life-sector from another, if one could not choose what of oneself to reveal here, what there. One underpinning of privacy claims, then, is the interest a person has in establishing, sustaining, and developing a personality. Taken off his guard, discovered in one role while projecting the persona of another, a person is embarrassed because he has lost control of his personae in an ambiguous situation and is at a loss to know what response is appropriate in terms of the complex notion he has of himself.

(Benn 1988: 282)

Nowadays, to most people, it comes as no surprise that it is beyond their power to control the flow of data related to them. Hence, user-driven content from social media may be merged with data from surveillance cameras, GP systems, apps, and data streams from the Internet of things, in which ambient intelligent systems in our environment provide personalized solutions to us, based on behavioral patterns revealed by sensor data. No wonder that we may easily find ourselves in situations in which data analytics could possible reveal unanticipated patterns of knowledge about our behavior and present an unfavorable picture of us. Hence, in our technology-mediated context, particularly attention is paid to *informational privacy* (Tavani 1999) as reflecting individuals' ability, or disability, to control the flow of personal information, including how information about them is exchanged and transferred.

Correspondingly, a justificatory conceptual framework can be found in Nissenbaum (2010), who has coined the term "contextual integrity" in order to explain for and tie adequate protection against informational moral wrongdoing. According to Nissenbaum, information flows always have to be seen in relation to context-sensitive norms, representing a function of: (1) the types of information in case, (2) the respective roles of communicators, and (3) principles for information distribution between the parties. Consequently, contextual integrity is defined, not as a right to control over information, but as a right to appropriate flows of personal information in contexts with right to two norms (Nissenbaum 2010: 127-ff.): norms of "appropriateness" and norms of "distribution," i.e., the moment of transfer of information from part X to $Y1 \ldots n$. Violations of one of these norms represent a privacy infringement.

Consequently, upholding privacy in the age of big data is challenging. As such, it is an open question whether anonymization techniques continue to represent feasible means for the protection of privacy (Barocas and Nissenbaum 2014). For instance, K-anonymity seeks to abstract data to a level, which hinders identification by hiding a unique individual among $k - 1$ others. Hence, the real data are transformed in a manner that makes re-identification impossible under a fixed anonymity threshold, which has been marked by a data controller. But, K-anonymity does not deliver privacy in cases where sensitive values in an equivalence class lack diversity. Here, examples of re-identification have been brought forward by Sweeney, who managed to point out the medical record of the governor of Massachusetts by means of data concerning age, health, and zip code, which for the given class were equivalent and thereby accidentally supported re-identification (Sweeney 2002). Similarly, an AOL user was identified once her AOL search record was linked to a photograph (Barboro and Zeller 2016). However, in this case, the applied anonymization technique was utterly simplistic as users' search queries were anonymized by assigning a unique number to them. Consequently, when AOL decided to release 20 million search queries, re-identification of users turned out to be a simple task—by simply following the unique number related to a given data trail, the user behind a search story could easily be identified.

Nevertheless, it may still be helpful to turn to advanced anonymization as a technique when seeking to reduce the risk of identifying data subjects, especially in cases in which personal data are of no interest in the first place. As an example, the ePOOLICE system architecture incorporated privacy safeguards in order to minimize the possibility for direct identification of data subjects and to further restrict the opportunity for indirect identification via aggregation of data fragments. Yet, new flows of information may cause a potential violation of contextual integrity, since information gathering via environmental scanning of communication streams

(Choo 1999) may well be judged inappropriate to that context and violate the ordinary governing norms of distribution within it. In that sense, the width spread scope of the scanning may raise concern among citizens. Although users are aware that data shared online are to some extent public, it is reasonable to assume that they probably do not expect their online activities to be made available as raw data sets for environmental scanning. For that reason, it is important to raise public awareness and to encourage transparency about technology-based crime fighting initiatives in order to promote democratic informed decisions that may guarantee the implementation of such systems in ways that are minimally invasive for human rights. As such, one of the main purposes of the European Union is to scaffold an environment of freedom, security, and justice, which also demands enhancing the capability of LEAs for carrying out efficient intelligence-led policing at a strategic level. Therefore, in order to balance societies' overall security needs without compromising citizens' right to privacy and freedom, different legal sources underscore the importance of guarding privacy against government intrusion. The ECHR recognizes the security of the individual as well as the right to privacy and data protection as fundamental rights on equal footing, and underscores that control by independent data authorities is an important part of European data protection (ECHR, article 8). Consequently, environmental scanning systems for strategic early warning have to be developed in legal compliance with EU law and member states privacy legislation. But, while such systems may slip under the radar of legal privacy restrictions, the scanning of public online sources may still imply privacy discomfort among people because of privacy concerns regarding information traffic across contexts representing distinctive spheres in life.

To sum up, individual citizens have gradually become more and more transparent to a variety of actors, and at the same time, they have experienced a reduction in transparency with right to knowledge of what is being known about them, where and by whom. On top of this, as Web users, we contribute to our own de-privatization by spreading information about ourselves on the Web or by enjoying the convenience of seamless Internet transactions based on personalized services in exchange for personal data. Needless to say, that the overall picture gives rise to lack of confidentiality and trust in relying on that intended or unintended information-based harm will not occur.

4 The Importance of Trust for a Flourishing Democratic Society

It is generally acknowledged that trust is essential for a flourishing society and that relations of trust are easily maintained and better preserved in communities with relatively low crime rates (Delhey and Newton 2003). At the same time, societal trust basically rests on the ability of citizens to rely on that in interacting with others, including governmental authorities, their integrity and autonomy will be respected (Nissenbaum 2010) and to provide for this, privacy is a highly held value, which has to be properly protected.

Hence, as citizens, we are reluctant toward any kinds of surveillance technologies, which may potentially restrict our privacy and thereby constrain our freedom. But privacy is not only important for individuals; privacy also has to be recognized as a societal good or collective value of crucial importance to economic and societal development in democratic liberal societies. In fact, one of the core goals of a democratic state is to scaffold individual liberty by protecting citizens against state interference. As pointed out above, in order to form our identity we must be capable of autonomous decision-making without interference from outside forces directing our choices (Kant 1785). If citizens fear intrusive agents of government in ordinary life contexts, they may start to adjust their actions in order not to contrast with mainstream behavior (Peissl 2003: 22; Regan 1995). An observation Reiman captures with reference to what he calls "informational panopticon" (Reiman 1995: 34), the consequences of which he elaborates on as follows:

> Trained by society to act conventionally at all times, people will come so to think and so to feel. Their inner lives will be impoverished to the extent that their outer lives are subject to observation. Infiltrated by social convention, their emotions and reactions will become simpler, safer, more predictable, less nuanced, more interchangeable.

(Reiman 1995: 41)

Within this kind of panoptic setting, creativity and drive in society may be hindered, if individuals feel an urge to carry out non-authentic performance-oriented behavior (Peissl 2003). As such, our capacity for situational awareness makes us able to adjust our behavior accordingly, to the given social context we find ourselves in, for better and for worse.

4.1 The Importance of Legitimacy and the Issue of Proportionality

In light of the above-mentioned reflections, the link between privacy and democracy cannot be emphasized too much, since privacy is an important condition for the flourishing of fundamental rights in liberal democracies, such as freedom of speech and information as well as citizens' freedom and opportunities for self-determination in matters regarding their civil and political life. This underscores the importance of citizens trusting that society will allow room for them to act as autonomous reasonable individuals. At the outset, citizens may trust that their governments will act to shield their interests and rights. However, societal trust and police legitimacy may come under pressure if the methods of law enforcement agencies gradually get detached from society and turn invisible to the public due to an increase in technology-led policing activities, on behalf of more traditional approaches with police presence in the communities. As such, surveillance initiatives by governmental organizations, in which one usually place trust, may infringe one's sense of dignity, which is a highly held ethical value reflected as an underlying basic principle in the formulation of both the ECHR and the charter of Fundamental Rights of the European Union (adopted in 2000 and entered into force in 2009) and explicitly stated in article one of the charter: "Human dignity is inviolable. It must be respected and protected."

Furthermore, even despite the presence of both legal and general public support to security technologies it is in principle never possible to establish with certainty whether the privacy interference caused by an environmental scanning system for crime fighting is proportionate to the end pursued, i.e., whether such a system is, in fact, a suitable, necessary, and adequate mean for fighting organized crime on a strategic level, since security advantages are not easy to calculate—neither ahead nor ex-post. Hence, from the fact that security technologies may have proved to be effective in the past, one cannot presuppose this outcome for systems in advance, especially not since environmental scanning systems for strategic early warning is targeted at long-term predictions of emerging future crime trends. Also, if it turns out to be the case, ex-post, that one observes a decline in organized crime after the implementation of such systems, it is not possible to validate that a given system by itself is a sufficient or necessary condition for causing this effect. It would demand a thorough evaluation to justify whether and how a system contributed to such an outcome. Consequently, one may have good reasons to presume that there is a causal relation, but one cannot know for certain that environmental scanning systems for early warning will represent a significant contribution in the fight against organized crime. Therefore, it is important that such systems are founded on robust privacy-preserving design frameworks.

5 Conclusion

In attempting to restrain organized crime and prevent formation of more resilient criminal systems, it is important that LEAs are equipped with tools enabling them to respond proactively to future crime challenges and safeguard public security. Here, big data-powered environmental scanning systems operating in tandem with human domain expertise, such as, for instance, the ePOOLICE system, may improve LEAs' capabilities for countering organized crime. Consequently, such systems may provide tools for strategic long-term planning, in which looking into the horizon can assist LEAs in the formation of hypotheses about future crime scenarios and emerging crime trends. In the context of such security technologies, one cannot exaggerate to much the importance of developing systems while applying a privacy-preserving design framework for balancing data privacy and data utility. In continuation thereof, once such systems are in place in LEA organizations, these efforts have to be accompanied by regularly privacy impact assessments and privacy awareness training of staff.

Tools of transparency are implemented in the current European data protection legislation, implying that it should be visible to individuals how their data are used in cases in which their privacy is traded off for state security by implementation of surveillance technologies. Even though environmental scanning systems for early warning are not surveillance technologies per se—they operate at the strategic level and do not aim at identifying individuals—it is still important to reach out to raise public awareness about such systems in order to facilitate informed decision-making and foster democratic dialog about their usage. Otherwise, citizens may form false beliefs and gradually lose trust in LEAs and government authorities and their abilities to protect individuals from informational wrongdoing (Van den Hoven 1997).

People gladly contribute to their own de-privatization, but at the same time, they fear government surveillance. The main reason for this attitude toward LEAs or governmental surveillance activities is of course that governments can put you to jail, and therefore it is of outmost importance to adhere to a precautionary principle when developing security technologies for LEAs—bearing in mind Benjamin Franklin's well-known quote: *Those who surrender freedom for security will not have, nor do they deserve, either one.*

Acknowledgements The author would like to thank Estelle De Marco, Henrik Legind Larsen, Raquel Pastor Pastor, and Javier Valls Prieto for valuable comments, which helped shape this chapter.

Research leading to these results has received funding from the European Union's Seventh Framework Programme (FP7/2007-2013) under grant agreement n° 312651.

References

Anderson, C. (2008). wired.com. *The end of theory: The data deluge makes the scientific method obsolete.* http://archive.wired.com/science/discoveries/magazine/16-07/pb_theory. Accessed February 8, 2017.
Ansoff, H. I. (1975). Managing strategic surprise by response to weak signals. *California Management Review, XVIII*(2), 21–33.
Barboro, M., & Zeller, T. (2016). A face exposed for AOL searcher no. 4417749. *The New York Times.* http://www.nytimes.com/2006/08/09/technology/09aol.html?pagewanted=all. Accessed February 8, 2017.
Barocas, S., & Nissenbaum, H. (2014). Big data's end run around anonymity and consent. In J. Lane, V. Stodden, S. Bender (Eds.), *Privacy, big data, and the public good* (pp. 44–75). New York: Cambridge University Press.
Benn, S. (1971). Privacy, freedom and respect for persons. In J. R. Pennock & J. W. Chapman (Eds.), *Privacy* (pp. 1–27). New York: Atherton Press.
Benn, S. I. (1988). *A theory of freedom.* New York: Cambridge University Press.
Biehn, N. (2013). The missing v's in big data: Viability and value. http://www.wired.com/insights/2013/05/the-missing-vs-in-big-data-viability-and-value/. Accessed February 8, 2017.
Boyd, D., & Crawford, K. (2012). Critical questions for big data. *Information, Communication & Society, 15*(5), 662–679.
Brewster, B., Polovina, S., Rankin, G., & Andrews, S. (2014). Knowledge management and human trafficking: Using conceptual knowledge representation, text analytics and open-source data to combat organized crime. In N. Hernandez, et al. (Eds.), *ICCS 2014, LNAI 8577* (pp. 104–117).

Callanan, C., Gercke, M., De Marco, E., & Dries-Ziekenheiner, H. (2009). Internet blocking—balancing cybercrime responses in democratic societies. http://www.aconite.com/sites/default/files/Internet_blocking_and_Democracy.pdf. Accessed February 8, 2017.

Choo, C. W. (1999). The art of scanning the environment. *Bulletin of the American Society for Information Science*, 21–24. Accessed February/March 1999.

De Marco, E. (2014). ePOOLICE, deliverables D3.3. *WP3—technical and legal/ethical constraints and system framework design*. https://www.epoolice.eu/EPOOLICE/servlet/document.listPublic. Accessed February 8, 2017.

Delhey, J., & Newton, K. (2003). Who trusts? The origins of social trust in seven societies. *European Societies, 5*(2), 93–137.

Floridi, L. (2012). Big data and their epistemological challenge. *Philosophy and Technology, 25*, 435–437.

Fried, C. (1968). Privacy: A moral analysis. *Yale Law Journal, 1*(77), 475–493.

Gerdes, A. (2014). A privacy preserving design framework in relation to an environmental scanning system for fighting organized crime. In K. Kimppa, D. Whitehouse, T. Kuusela, & J. Phahlamohlaka (Eds.), *ICT and society: 11th IFIP TC9 International Conference on Human Choice and Computers, HCC11, 2014* (pp. 226–239). Turku, Finland: Springer. July 30–August 1, 2014.

Hilbert, M. (2016). Big data for development: A review of promises and challenges. *Development Policy Review, 34*(1), 135–174.

IOCTA. (2015). Internet organized crime threat assessment 2015. https://www.europol.europa.eu/content/internet-organised-crime-threat-assessment-iocta-2015. Accessed February 8, 2017.

Johnson, D. (1994). *Computer ethics*. Prentice Hall.

Jonas, J., & Harper, J. (2006). *Effective counterterrorism and the limited role of predictive data mining* (pp. 1–11). Policy analysis no. 584. December 11, 2006.

Kant, I. (1785). *Grundlegung zur Metaphysik der Sitten* (Akademiausgabe, vol. IV). http://www.korpora.org/Kant/aa04/392.html. Accessed February 8, 2017.

Kitchin, R. (2014). Big data, new epistemologies and paradigm shifts. *Big Data & Society*, 1–12.

Laney, D. (2001). Data management: Controlling data volume, velocity and variety. http://blogs.gartner.com/doug-laney/files/2012/01/ad949-3D-Data-Management-Controlling-Data-Volume-Velocity-and-Variety.pdf. Accessed February 8, 2017.

Liotta, P. H. (2002). Boomerang effect: The convergence of national and human security. *Security Dialogue © 2002 PRIO, 33*(4), 473–488. (SAGE Publications).

Mayer-Schönberger, V., & Cukier, K. (2013). *Big data: A revolution that will transform how we live, Work and think*. USA: John Murray.

Nissenbaum, H. (2010). *Privacy in context—technology, policy and the integrity of social life*. Stanford: Stanford Law Books.

PCAST. (2014). *Report to the President—big data and privacy: A technological perspective*. Executive Office of the President's Council of Advisors on Science and Technology. https://www.whitehouse.gov/sites/default/files/microsites/ostp/PCAST/pcast_big_data_and_privacy_-_may_2014.pdf. Accessed February 8, 2017.

Peissl, W. (2003). Surveillance and security: A dodgy relationship. *Journal of Contingencies and Crises Management, 1*(11), 19–24.

Perry, W. L., McInnis, B., Price, C. C., Smith, S. C., & Hollywood, J. S. (2013). Predictive policing. The role of crime forecasting in law enforcement operation. *RAND corporation, safety and justice program*. RAND Corporation. http://www.rand.org/pubs/research_reports/RR233.html. Accessed April 25, 2016.

Rachels, J. (1975). Why privacy is important. *Philosophy & Public Affairs, 4*(4), 323–333.

Raguse, M., Meints, M., Langfeldt, O., & Peissl, W. (2008). Prepatory action on the enhancement of the European industrial potential in the field of Security research. *Technical report, PRISE*. http://www.prise.oeaw.ac.at/publications.htm. Accessed April 25, 2016.

Ratcliffe, J. H. (2011). *Intelligence-led policing*. New York, USA: Routledge.

Regan, P. M. (1995). *Legislating privacy: Technology, social values, and public policy*. USA: The University of North Carolina Press.

Reiman, J. H. (1995). Driving to the panopticon: A philosophical exporation of the risks to privacy posed by the highway technology of the future. *Santa Clara High Technology Law Journal, 11* (1), 27–44.

Schroeck, M., Shokley, R., Smart, J., Romero-Morales, D., & Tufano, P. (2012). Analytics: The real-world use of big data—how innovative enterprises extract value from uncertain data. *Executive Report—IBM Global Business Service*, 1–20.

SOCTA. (2013). *Serious organized crime threat assessment 2015*. https://www.europol.europa.eu/latest_publications/31. Accessed February 8, 2017.

Sweeney, L. (2002). K-anonymity: A model for protecting privacy. *International Journal of Uncertainty Fuzziness Knowledge Based Systems, 10,* 557–570.

Tavani, H. (1999). Informational privacy, data mining, and the internet. *Ethics and Information Technology, 1,* 137–145.

Van den Hoven, J. (1997). Privacy and the varieties of informational wrongdoing. *Computers and Society*, 33–37.

Horizon Scanning for Law Enforcement Agencies: Identifying Factors Driving the Future of Organized Crime

Timothy Ingle and Andrew Staniforth

1 Introduction

Contemporary crime is significantly challenging the capacity and capability of law enforcement agencies (LEAs) across the world. From terrorism to cybercrime, drug trafficking to child sexual exploitation, criminal investigations being conducted today have a far-reaching transnational dimension which substantially increases their complexity beyond similar crime types committed less than a decade ago. According to Europol, there are an estimated 3600 organized crime groups (OCGs) active in the European Union (EU).[1] These groups are becoming increasingly networked in their organization and behavior, characterized by a group leadership approach and flexible hierarchies. Through advances in international trade, an ever-expanding global transport infrastructure and the rise of the Internet and mobile communication have all served to engender a more international and interconnected form of serious and organized crime. As a result, there is an increased tendency for OCGs to cooperate with or incorporate into their membership a greater variety of nationalities. This phenomenon has resulted in an increased number of heterogeneous groups that are no longer defined by nationality or ethnicity, which provides evidence of how serious and organized crime is being fundamentally affected by the process of globalization. Contemporary criminals act undeterred by geographic boundaries and can no longer be easily associated with specific regions or centers of gravity. Despite this, ethnic kinship, linguistic and historic ties still remain important factors for building bonds and trust between criminals, and these factors often determine the composition of the core OCGs.

---

[1]Europol *Organised Crime Threat Assessment Report* (2015).

T. Ingle · A. Staniforth (✉)
West Yorkshire Police, Wakefield, England, UK
e-mail: andrew.staniforth1@westyorkshire.pnn.police.uk

As criminal organizations diversify their activities and increase their collaboration at both the European and global level, it is essential that LEAs have the capability to tackle modern crime. An important aspect of the fight against organized crime is the ability to detect and disrupt OCGs through proactive and preventative interventions. This approach requires LEAs to improve ways in which they can identify early signs, signals and indicators that warn them of the emergence of OCG criminality. To develop new ways in which to combat contemporary organized crime, and to work together to provide a solution to effectively address this operational challenge, during 2012–2015 leading LEAs formed a network of organized crime security professionals from across the EU, including Europol, Guardia Civil, FHVR in Germany and West Yorkshire Police in the UK, joining forces with academic institutions, government departments and innovators from private industry as part of a multi-disciplinary consortium to progress ePOOLICE (early Pursuit against Organized crime using environmental scanning, the Law and Intelligence systems),[2] a three-year research and innovation project funded by the Framework Programme of the European Commission Executive Research Agency. The primary purpose of this project was to build an efficient environmental scanning system to enhance early warning from open sources of information (OSINF) for LEAs.

The purpose of this chapter is to articulate the methodology used to identify factors driving future crime designed to inform the development of ePOOLICE and to provide the rationale for selecting certain crime types for scenario development to test and validate tools. An essential element of this chapter serves to contextualize the contemporary operational landscape from an LEA practitioner perspective, describing the current intelligence discipline into which horizon scanning for open sources of information (OSINF to OSINT) is fast becoming an integral part of the prevention, investigation and detection of organized crime.

2 Methodological Approach

The ePOOLICE project network of organized crime security professionals, together with associated end-users from across the EU, constructed a number of case studies to meet their objective of refining a methodology for monitoring heterogeneous information sources. This was done in order to develop an indicator list which would then become a key mechanism to identify changes in 'ordinary' or 'common' situations, the interpretation of which could lead to the classification of the changes as at least potentially being representative of weak signals of emerging crime threats. To ensure the development of the case studies that were aligned to the overarching strategic object of ePOOLICE, which was to develop an efficient and

---

[2]https://www.epoolice.eu/.

effective prototype environmental scanning system for the detection of emerging organized crime threats, the network of organized crime security professionals focused on the following:

- understanding the challenge;
- responding to the challenge;
- defining open-source intelligence (OSINT);
- considering available intelligence sources;
- identifying trends when scanning the environment; and
- analyzing factors that affect crime.

The above work was accomplished through the development of:

- end-user needs and requirements;
- the development of real and representative scenarios;
- evaluating and assessing factors affecting new and emerging organized crime; and
- end-user evaluation.

The general consensus among the diverse range of organized crime security professionals was that, in terms of understanding the operational landscape, it was important to reflect upon a number of factors, such as how rapid global development and advances in technological changes, together with the flexibility, adeptness and resourcefulness of OCGs to embrace and adopt new technologies as they equally evolve at pace, should be best considered. Through the Internet, smart mobile communications and an ever-expanding global transport infrastructure, organized crime networks are now international in scope and have no geographic or ethnic boundary. While it is recognized that the global economic crisis of 2008 has led to shifts in criminal markets, OCGs have maximized these opportunities to generate profit by distributing illicit commodities. Many OCGs have utilized their existing infrastructure to expand into other forms of criminality, exploiting the reduced consumer spend by venturing into the counterfeited goods market of luxury and everyday items.

Responding to meet the scale and complexity of the identified operational challenges of contemporary organized crime remains the highest priority of governments across the world in their first duty to provide safety and security for their citizens. In order to meet these challenges effectively, LEAs need not only to amplify and accelerate their uptake and effective use of new communications technologies to catch up with the increasingly sophisticated organized criminal, but to combat contemporary crime fresh efforts and a real focus upon developing strategic early warning capabilities is considered essential.

The development of an effective and efficient early warning system for LEAs is heavily dependent upon environmental scanning (ES). ES is the process of continually acquiring information on events occurring globally to identify and interpret potential trends. Providing an explanation for the trend and assessing its implications are also an important part of ES for LEAs. To progress the ePOOLICE project, the networks of organized crime security professionals, together with legal experts, developed a prototype of an ES tool. In developing the ES tool, a number of key objectives were set which included:

- To conduct research into technologically, actor-driven systems and tools which support environmental scanning to enable the rapid identification and qualification of new organized crime threats.
- To scan the environment to feed new and emerging threats into the serious and organized crime threat assessment processes.
- To identify a combination of technological resources and human actors that serves to improve the process of detecting and selecting new OC threats that warrant EU level analysis and EU-wide responses.

To build an effective ES tool, which relies upon the receipt and analysis of open-source intelligence (OSINT) to determine and detect crime trends and patterns emerging across the globe, a clear understanding of OSINT was required, as well as the position of OSINT within the current suite of LEA intelligence collection disciplines.

3 Open-Source Intelligence

Since before the advent of advanced technological means of gathering information, as part of their overall all-source-based criminal intelligence work, LEAs have planned, prepared, collected and produced intelligence from publicly available information and open sources (OSINF to OSINT) to gain knowledge and understanding in support of preventing crime and pursuing criminals.[3] While traditional threats from crime have historically begun at the most local level, in today's increasingly interconnected and interdependent world, many new hazards have a cross-border, transnational dimension, being amplified by the Internet, online social networks and smarter mobile communications. Social and technical innovations are now occurring at an ever-increasing speed, causing fast and drastic changes to society. These changes, driven by the possibilities offered by new and emerging technologies (as introduced above), affect citizens, their wider communities, the private sector, the government and, of course, the police.

To effectively tackle the full spectrum of contemporary security hazards, LEAs have tapped into an increasingly rich source of intelligence that is gathered from publicly available information. The relentless pursuit of intelligence by police officers to keep communities safe via the use of open sources of information has produced the fastest growing policing discipline of the twenty-first century—a

---

[3]See, for instance, as discussed throughout Ratcliffe (2016).

discipline which adds significant value and increasing efficiency to combating contemporary crime, called OSINT.

Various kinds of intelligence—military, political, economic, social, environmental, health and cultural—provide important information for decisions. Intelligence is very often incorrectly assumed to be just gathered through secret or covert means, and while some intelligence is indeed collected through clandestine operations and is known only at the highest levels of government, other intelligence consists of information that is widely available. According to the Federal Bureau of Investigation (FBI) in the USA, and recognized by many LEAs across the world operating today, there are various ways of gathering intelligence that are collectively and commonly referred to as 'intelligence collection disciplines'[4] which are shown in Table 1.

Within this suite of specialist intelligence collection disciplines, OSINT has now found its place, being extensively used by local and national LEAs, intelligence agencies and the military. Given the scale, accessibility and high yield of intelligence return for minimum resource, OSINT provides the glue which binds, compliments and increasingly corroborates and confirms other LEA intelligence functions, all of which are relevant operational reasons as to why OSINT has quickly become a rich source of information to disrupt and detect the modern criminal.

According to the FBI, OSINT is the intelligence discipline that pertains to intelligence produced from 'publicly available information that is collected, exploited and disseminated in a timely manner to an appropriate audience for the purpose of addressing a specific intelligence and information requirement' (See Footnote 4). OSINT also applies to the intelligence produced by that discipline. OSINT is also intelligence developed from the overt collection and analysis of publicly available, and OSINF is not under the direct control of government authorities. OSINT is derived from the systematic collection, processing and analysis of publicly available information in response to intelligence requirements. Two important related terms are 'open source' and 'publicly available information,' which are defined by the FBI as:

- Open source is any person or group that provides information without the expectation of privacy—the information, the relationship, or both is not protected against public disclosure. Open-source information can be publicly available but not all publicly available information is open source.
- Publicly available information is data, facts, instructions or other material published or broadcast for general public consumption; available on request to a member of the general public; lawfully seen or heard by any casual observer; or made available at a meeting open to the general public (See Footnote 4).

---

[4]Federal Bureau of Investigation, Intelligence Branch, Intelligence Collection Disciplines https://www.fbi.gov/about-us/intelligence/disciplines.

Table 1 Intelligence collection disciplines

Discipline	Description
Human Intelligence (HUMINT)	The collection of information from human sources. The collection may be conducted openly, by police officers interviewing witnesses or suspects during the general course of their duties, or it may be collected through planned, targeted, clandestine or covert means. A person who provides information to the police as part of a covert relationship is known as an 'agent' or 'source' and in the UK is known as a Covert Human Intelligence Source (CHIS). LEAs who exploit covert sources of HUMINT have specialist and dedicated units with trained officers given the increasing levels of risk, security, complexity and legal and ethical considerations and compliance required
Signals intelligence (SIGINT)	Refers to electronic transmissions that can be collected by ships, planes, ground sites or satellites. More commonly used in higher policing operations or investigations which pose potential threats to national security. SIGINT is generally conducted by, and the responsibility of the military or intelligence services rather than domestic LEAs, although at federal and national levels SIGINT may be used by LEAs in support of serious and organized crime and terrorism investigations which threaten public safety
Communications intelligence (COMINT)	A type of SIGINT and refers to the interception of communications between two or more parties. This can be conducted in real-time or captured and stored for future interrogation. SIGINT is primarily the responsibility of intelligence agencies such as the National Security Agency (NSA) in the USA or the Government Communications Headquarters (GCHQ) in the UK. As with SIGINT, COMINT may be used by LEAs in support of organized crime and terrorism investigations which presents a serious threat to public safety
Imagery intelligence (IMINT)	Also referred to as photo intelligence (PHOTINT), IMINT includes aerial and more increasingly satellite images associated with reconnaissance. Traditionally the preserve of military and intelligence agencies, IMINT is increasingly used by LEAs being captured by their aerial capabilities including helicopters and more recently unmanned aerial vehicles (UAVs)
Measurement and signatures intelligence (MASINT)	A relatively little-known collection discipline that concerns weapons capabilities and industrial activities. MASINT includes the advanced processing and use of data gathered from overhead and airborne IMINT and SIGINT collection systems
Telemetry intelligence (TELINT)	Used by the military and intelligence agencies, sometimes deployed to indicate data relayed by weapons during tests in the international monitoring to counter the proliferation of chemical, biological, nuclear or radiological weapons

(continued)

Table 1 (continued)

Discipline	Description
Electronic intelligence (ELINT)	Indicates electronic emissions picked up from modern weapons and tracking systems. Both TELINT and ELINT can be types of SIGINT and contribute to MASINT as part of a multi-pronged approach to intelligence gathering operations from the military and intelligence agencies

OSINT collection by LEAs is usually accomplished through monitoring, data-mining and research. OSINT refers to a broad array of information and sources that are generally available, including information obtained from the media (newspapers, radio, television, etc.), professional and academic records (papers, conferences, professional associations, etc.) and public data (government reports, demographics, hearings, speeches, etc.). OSINT includes Internet online communities and user-generated content such as social-networking sites, video- and photo-sharing sites, wikis and blogs. It also includes geospatial information such as hard- and softcopy maps, atlases, gazetteers, port plans, gravity data, aeronautical data, navigation data, geodetic data, human terrain data, environmental data and commercial imagery. Much is offered to LEAs from OSINT.

4 Identifying Trends

To progress the identification of crime trends to support and compliment the OSINT and ES elements of the ePOOLICE tool, the network of organized crime professionals used the Political, Economic, Societal, Technological, Legal and Environmental (PESTLE) methodology. Widely used as an analysis framework, PESTLE is a concept in marketing principles, used by private business to track the environment they are operating in or planning to launch new services, products or services. The PESTLE concept naturally lends itself to scanning the environment for changes, which may be a trigger for change in crime types, criminal markets and networks. Utilizing this analysis model provided the potential for a rich and diverse set of indicators which would provide an excellent and holistic view of the global environment. The analysis model ensures that consideration of all political, economic, social, technical and environmental issues of potential change was considered, all of which are potential identifiers for new and emerging threats. Although the importance of legal issues for factors driving changes in organized crime was considered, for the purposes of the development of the ES, the legal aspects did not fall within the scope of the project objectives given the sheer volume and complexity of identifying each nation's legal frameworks. To refine PESTLE to meet the needs of ePOOLICE, the network of organized crime security professionals considered the following series of questions:

- What is the political situation of a country and how can this affect crime?
- What are the prevalent economic factors?
- What impact does the change in culture have on crime?
- What technological innovations are likely to appear and what challenges will these present?
- Is current legislation and policy fit for purpose, does it cater for anticipated change, and is it able to keep pace with evolving technology?
- What part does the environment play in the changes in criminality?
- Does the global population consider social responsibility?

The answers to these questions were captured and are detailed in the following sections.

4.1 Technology Factors

As technology advances at an alarming pace, the consumer is provided with limitless services and functionality, which enriches life providing convenience and connection. However, these advancements offer OCGs overwhelming possibilities providing them with anonymity through online criminal marketplaces accessed through anonymized networks and global associations as geographic boundaries no longer exist. The development of Crypto Currencies and the ability to commit crime in a virtual space with no physical connection to a scene or a victim are also key technological factors, as illustrated in Fig. 1.

4.2 Political Factors

As political regimes fall and fracture, and international cooperation fails, the migration of millions of people continues to escalate at an unprecedented scale providing opportunities for OCGs to exploit citizen vulnerabilities. The identification, detection and prosecution of criminals, especially those operating within close proximity to areas of conflict or those governed by oppressive regimes, are thwarted by the sheer scale of migration and the unknown population within a country's geographic borders. The lack of strong cooperation and effective coordination between nations through sanctions, political discourses of the mass movement of people and legal wrangling over border controls further exacerbates the situation through the denial of information sharing and joint cooperation allowing OCGs to thrive. The political factors providing an opportunity for organized crime are shown in Fig. 2.

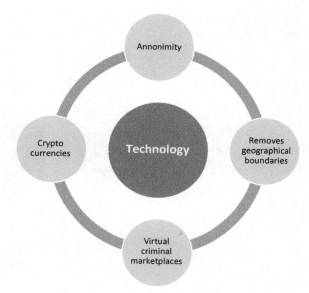

Fig. 1 Technological factors and trends driving future crime

Fig. 2 Political factors and trends driving future crime

4.3 Economic Factors

As economies shrink, the provision of services to combat crime reduces providing an environment appealing to the OCG. Poverty drives the need and marketplace for counterfeit goods providing a growing economy for illicit products fueling the need

Fig. 3 Economic factors and trends driving future crime

and opportunity for expanding OCGs who diversify to seize this opportunity. As economic buoyancy shifts around the globe, the OCGs follow the trail to increase their wealth through fraud and diversification into other crime types. The economic factors driving crime are illustrated in Fig. 3.

4.4 Social Factors

As demographics change and population density increases, communities who quickly form are vulnerable through a lack of identity, unity and social cohesion. Populations no longer have affiliation to the state and the acceptance of law and order, and social responsibility is diminished leading to the acceptance and tolerance of crime, trade in illicit goods and the erosion of civic duty to report crime and support the police and their partners in efforts to protect the public. The social factors driving the future of crime are shown in Fig. 4.

4.5 Environmental Factors

The location and timing of environmental disasters are primarily unpredictable events and can provide devastation to large populations. Whether man-made or purely natural occurring events, these factors destroy lives and communities fueling

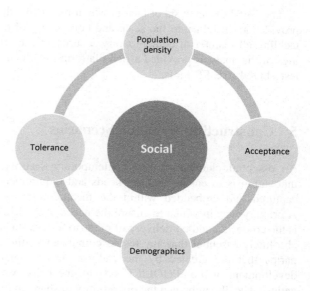

Fig. 4 Social factors and trends driving future crime

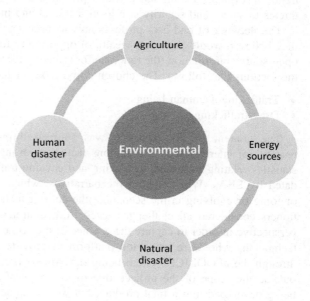

Fig. 5 Environmental factors and trends driving future crime

migration, international dispute, lack of trust in government response, death and injury leading to vulnerability providing unique opportunities for the introduction of criminality. The environmental factors impacting upon future crime are shown in Fig. 5.

The PESTLE analysis of factors which may drive the future of organized crime provides an insight into the scale and complexity of the challenges LEAs face of tackling all manner of contemporary security threats, risks and hazards. The findings of the analysis provided a rich platform on which to construct mechanisms to test ePOOLICE ES tools.

5 Constructing Dynamic Scenarios

An essential element of any contemporary security research, development or innovation is to ensure that the needs and requirements of the LEA end-user are captured and embedded within the methodology, design and delivery phases. A primary way in which to ensure the needs of the LEA are designed into research is the co-design and co-delivery of pilots to test, evaluate and validate progress. It is absolutely essential that any test is completed within the context of realistic scenarios that are created by the end-user professional and practitioner. For the development of the ePOOLICE system, the LEAs were actively encouraged and enthusiastically embraced the opportunity to share and provide current and ongoing trends in organized criminality which provided acute operational challenges and threats to safety and security at a local, national and international level.

The network of end-user professionals focused on two key crime types which they believed would provide a wealth of opportunity for collecting and processing open-source information (the rationale for selecting these crime types is provided in the sections that follow). The chosen crime types included:

- Traffickingof human beings
- Drug trafficking

The content of the scenarios was sourced through extensive open-source research of international organizations factual reporting on criminality, trends and statistics relating to human trafficking and cocaine trafficking, and was later validated by LEAs. When writing the narratives, which were all based on historic, ongoing or evolving crime scenarios, the subject matter experts and LEA practitioners considered all challenges faced from a strategic, tactical and operational perspective in order to capture the essence of the crime and provide context for the technicians who were using the platform to provide the technological solutions through the ePOOLICE system being developed. These challenges were at times outside the scope of the project, due to its 'proof-of-concept' nature rather than being developed into a final product, but were included to provide a deep insight into the investigation of organized crime. The scenarios created were developed according to the LEAs requirements and priorities in order to validate the real-world performance of outcomes of the project. LEAs were encouraged to provide case studies so that taken together, they provided sufficient cross-referencing for the technicians to start building the software for an effective ES tool to assist the early pursuit of emerging organized crime trends.

To provide a further layer of operational credibility to the scenarios, ensuring they had 'real-world' lifelike substance and would stand up to scrutiny from LEAs across the EU and beyond, an investigative framework was adopted to challenge and test the scenarios utilizing proven techniques, while asking searching questions in order that the research and data collection provided a representative scenario that had validity was realistic and would be useful to technicians creating a knowledge repository and the scanning system. It was also considered essential to ensure realistic scenarios were created and that all engaged in the development of the scenarios increased their level of awareness and understanding of crime. As a direct result, a number of factors were considered that are known to affect the volume and type of crime. These were as follows:

- Population density
- Composition of the population (youth concentration)
- Economic conditions, including income, poverty level and job availability
- Cultural factors and educational, recreational and religious characteristics
- Community cohesiveness and demographics
- Climate
- Effectiveness of law enforcement agencies and partners
- Administrative and investigative emphases of law enforcement
- Criminal justice system (robust and effective)
- Culture of crime-reporting practices
- Social acceptance and tolerance

When developing the scenarios, consideration was also given to the means, motive and opportunity of the OCG within the scenario, which included:

- **Means**—what 'capability' does the organized crime group, victim, witness or accessory have?
- **Motive**—what is the 'mind-set' of the organized crime group, victim, witness or accessory?
- **Opportunity**—what 'planning' has commenced by the organized crime group, victim, witness or accessory?

These key areas were scrutinized through questioning every event or situation through a set of interrogative pronouns, commonly known in the law enforcement field as the '5 × WH + H' method—Who, What, When, Where + How (Cook and Blackstone 2015). The 5 WH model constitutes a formula for capturing a holistic version of events. According to the principles of this model, a report can only be considered complete if it answers these questions starting with an interrogative word. Therefore, in each element of the scenario, we asked:

- **Who** is the victim?—Victim details and why this victim?
- **What** happened?—Precise details on incident/occurrence
- **When** did it happen?—Temporal issues such as relevant times

- **Where** did it happen?—Geographic locations, national/international?
- **Why** did it happen?—Motivation for crime
- **How** did it happen?—Precise modus operandi details

While considering these questions at each stage of the process, the network of organized crime security professionals considered in what form and from what source the answers may be available, which included:

- **Technical**—Communication methods, media, CCTV, financial transactions
- **Human**—Victim, witness, suspect accounts
- **Physical**—ID, travel, documents, purchases, money, property, reports

Constructing the dynamic scenarios was an important element to ensure the ePOOLICE tool could be stress-tested and validated via a realistic type of crime and case. Selecting the types of crime to provide the framework for the scenarios required further assessment and considered selection.

6 Selected Crime Types Rationale

The two chosen crime types of drug and human trafficking were appealing for a number of reasons as they presented a wealth of opportunity for data collection at various stages of the criminal process. In each case, the commodity, whether it be a person or a drug, commenced a journey ending in exploitation or a sale. This led to the generation of 'cash,' which, in turn, becomes a product of the criminal enterprise that enters a process to be laundered and realized by the OCG. These journeys provided an excellent basis to create a story-based narrative, which explored each stage of the criminal journey of the product or person revealing signs and indicators from a wealth of different sources. These early signs, indicators and weak signals were of significant interest to the ePOOLICE team. In the construction of the scenarios, each stage of the criminal journey was explored considering visible signs and indicators. Figures 6 and 7 show process diagrams for drug and human trafficking, which were constructed to identify the major stages of the criminal journey.

The scenarios to test, evaluate and validate the technical tools and scanning system of ePOOLICE were grounded in the operational realities of the LEA practitioner's experiences. While the far-reaching and long-term impact of the trafficking, importation and use of illegal drugs on individuals and wider society is well documented and understood by academia, governments and LEAs, the extent of human trafficking has yet to capture the full attention of all policy-makers and police professionals at the most local level. Despite Article 4 of the 1948 Universal Declaration of Human Rights stating that people should not be bought and sold, the trafficking of human beings is the fastest growing global crime (Chalke 2009). According to Europol, the total profit from the global criminal enterprise of human trafficking is second only to the trafficking of drugs (Chalke 2009). The evidence provided by Europol confirmed and validated the decision of the ePOOLICE team

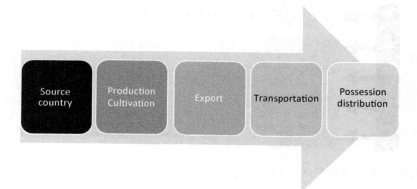

Fig. 6 Process diagram: criminal journey of drug trafficking

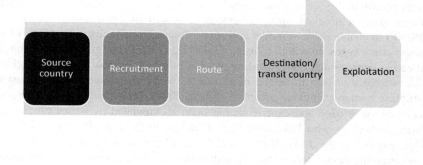

Fig. 7 Process diagram: criminal journey of human trafficking

to focus upon drug and human trafficking. To provide some sense of the scale and scope of human trafficking, it is estimated by UNICEF that human trafficking generates between 10 and 12 billion dollars every year, which includes criminal profit from an estimated 1.2 million children being trafficked every year, the equivalent of two children every minute trafficked for the purposes of sexual exploitation (Chalke 2009). Moreover, the United States State Department has revealed the wider health and security concerns of human sex trafficking, suggesting that: 'The health implications of sex trafficking extend not only to its victims, but also to the general public. Sex trafficking is aiding the global dispersion of HIV subtypes (Getu 2006).' The trafficking and slavery of human beings is an international organized crime, with the exploitation of people at its heart. It is an abuse of basic rights, with organized criminals preying on vulnerable people to make money (Burns-Williamson 2016). Behind the statistics of people trafficked

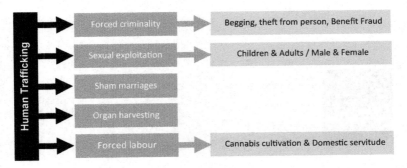

Fig. 8 Component parts of human trafficking criminality

are the very real victims. They are mothers and fathers, husbands and wives, sons and daughters, brothers and sisters, all torn apart and their lives ruined by abuse for financial gain. Substantial interest remains in addressing these issues.

To capture the personal impact of human trafficking, as well as the complex layers of criminality on a transnational level, the organized crime security professionals group embarked upon an extensive search of open-source information (OSINF) to enable them to model and create dynamic and realistic scenarios of the chosen crime types. In addition, using OSINF to OSINT, the group drew upon the collective learning and understanding of LEA efforts to combat human trafficking. LEA strategies proven to be the most successful are those that have emphasized disrupting human trafficking activities (Zhang 2007). Over the last decade or so, a wealth of knowledge and experience has been accumulated by police officers and border authorities who have investigated human trafficking offenses and OCGs. While engaged in the data collection and the development of the scenario, the group considered the parameters of human trafficking and the opportunities it presented. As an 'umbrella' crime, human trafficking exploits its victims through a number of different crime types. By utilizing this pan-European emerging and expanding category of crime, it was possible to create a scenario which considered a number of crime types and a diverse range of data sets. Figure 8 illustrates the diverse typology of crimes within human trafficking and the totality of threat, risk and harm to victims and the communities in which they operate. Figure 8 also reveals the complexity of investigating and detecting human trafficking, as well as the impact upon finite LEA resources.

When researching and providing dynamic scenarios, the ePOOLICE team considered the identified PESTLE factors and identified associated 'push' and 'pull' factors. For human trafficking, these were factors that were critical in either pushing people toward or increasing their vulnerability to becoming a victim of human trafficking, and those factors that pulled them toward the criminal traffickers that were being used to coerce and exploit victims. Table 2 illustrates the 'push' and

Table 2 Identified 'push' and 'pull' factors of human trafficking

Identified 'pull' factors of human trafficking	Identified 'push' factors of human trafficking
• Expectation of employment and financial reward • In the sex industry, expectation of rich financial rewards • Access to material benefits associated with the West • Improved social position and treatment • Perceived 'glamor' of Western European daily life • Demand of the 'exotic' women as prostitutes • Employers' demand of cheap labor • Consumers' demand of cheap goods and services • Relatives and friends living in the destination country • Returning migrants, legal and illegal, who have allegedly made a better living for themselves?	• Poverty—and the feminization of poverty • Lack of education • Unemployment • Gender discrimination • Domestic violence • Life within dysfunctional families • Impact of conflict or transition on countries • Lack of opportunity for legal migration • Lack of opportunity or alternatives

'pull' factors that were identified for human trafficking and used to enrich the human trafficking scenario designed to inform, build and test the ePOOLICE tools.

Taken together, and due to all their features as discussed, the careful and considered selection of human trafficking and drugs trafficking as crime types provided an excellent platform upon which to construct complex scenarios to test and validate the ePOOLICE tool.

7 Conclusion

The world is being reinvented by open sources. Publicly available information can be used by a variety of individuals, and organizations to expand a broad spectrum of objectives and LEAs are increasingly making effective use of this free and accessible source of information. While the Internet and online social networks have positively enriched societal communications and economic opportunities, these technological advancements have changed—and continue to change—the very nature of organized crime, serving to breed a new sophisticated and technically capable criminal.

Significant changes in organized crime investigative practices have also been made. Harnessing the power of OSINF to OSINT via ES capabilities created by tools such as those developed within the ePOOLICE project presents a unique opportunity for governments to address the increasing threats from the international reach of organized crime at relatively low cost. But the implementation and execution of ES tools raises acute concerns for existing data storage capacity, together with the ability to share and analyze large volumes of data and to positively act upon the information discovered. LEAs must also ensure that their access and use of publicly available information through ES tools is within national and international legal frameworks, which provides legitimacy and confidence that ES which exploits OSINT is being used appropriately. The accessibility of OSINT and the introduction of ES capabilities will also require the rigorous review and potential overhaul of existing intelligence models and associated processes to ensure all in authority are ready to exploit the operational advantages that ES of OSINT can yield.

An important lesson learned from LEA engagement, contribution and active participation in the ePOOLICE project is that, in order to build effective tools to counter organized crime, LEAs must co-design, co-deliver and co-produce research, development and innovation activity. LEAs working with other government departments, academia and private industry are stronger and represent a more powerful agent for positive change. Yet despite an increasing number of LEAs engaging in research, they are acutely aware that not only can they prevent crime on their own but that it remains regrettably unlikely that they will not prevent all new and emerging types of organized crime which can go on to ruin the lives of many citizens. In the light of that conclusion, all in authority must dedicate themselves to increasing LEA capabilities and developing new approaches to better protect the public they serve. To ignore or dismiss the positive benefits of OSINT and ES would therefore be both misplaced and unwise, as all citizens expect the LEAs and their intelligence agency counterparts to take the necessary steps to keep them safe. Harnessing the power of ES is fast becoming a game changer for LEA policy-makers, professionals and practitioners. Given the scale and complexity of the threats from contemporary organized crime, and the drivers of future crime articulated in this chapter, LEAs must continue to advance counter measures to keep us all safe and, most importantly, seek new ways in which to embed progressive developments to ensure they provide themselves every opportunity to detect crime and bring contemporary organized criminals to justice.

References

Burns-Williamson, M. (2016). *Encouraging innovation in policing—Tackling slavery and exploitation*. UK: Secure Societies Institute, University of Huddersfield.
Chalke, S. (2009). *Stop the traffik—people shouldn't be bought and sold*. Oxford: Lion Hudson.
Cook, T., & Tattersall, M. (2015). *Blackstone's handbook of criminal investigation*. Oxford: Oxford University Press.
Getu, M. (2006). *Human trafficking and development: The role of microfinance in transformation* (Vol. 23, p. 3). United States: State Department.
Ratcliffe, J. H. (2016). *Intelligence-led policing* (2ed., p. 216). London: Routledge (esp. col. 2).
Zhang, S. X. (2007). *Smuggling and trafficking in human beings*. Westport, United States: Praeger Publishers.

Macro-environmental Factors Driving Organised Crime

José María Blanco and Jéssica Cohen

> Before you become too entranced with gorgeous gadgets and mesmerizing video displays, let me remind you that information is not knowledge, knowledge is not wisdom, and wisdom is not foresight. Each grows out of the other, and we need them all.
> Arthur C. Clarke

1 Organised Crime in VUCA (Volatility, Uncertainty, Complexity and Ambiguity) Times

The world is becoming more complex, chaotic and volatile. The way we understand the challenges posed by the new scenarios will determine our ability to address them on a theoretical way but also in the physical environment.

In the 1990s, the US military outlined what would be a new military training programme focused on the emerging threats. Its parameters were defined with a clear objective: to develop the capacity of its forces to act under contexts of high complexity. A new need arose after identifying the main characteristics that would determine future scenarios known as VUCA (acronym for volatility, uncertainty, complexity and ambiguity) environments. As a result of this initiative, in 2004 the first results of a new training programme known as *"Think like a Commander"* (TLAC) were published. Its conclusions were defined clearly on the first lines of the document: *"Success in future operations depends on the ability of leaders and soldiers to think creatively, decide promptly, exploit technology, adapt easily, and act as a team"*.

J.M. Blanco (✉)
Centre of Analysis and Foresight, Guardia Civil, Madrid, Spain
e-mail: jmblanco@guardiacivil.es

J. Cohen
International Security Intelligence Analyst, Private Sector, Madrid, Spain
e-mail: mail@jessicacohen.es

© Springer International Publishing AG 2017
H.L. Larsen et al. (eds.), *Using Open Data to Detect Organized Crime Threats*, DOI 10.1007/978-3-319-52703-1_7

Regardless of its military use, these variables define a new environment to be considered in different disciplines and sectors, in order to manage the future. A strategy cannot exist if there is not future thinking.

1.1 Volatility (Also Called Turbulence)

Changes come fast, with growing difficulties to predict them and identify trends or patterns. This situation generates instability. But we must tackle these changes, developing a continuous adaptive framework and a quick response process. In response to the Darwin legacy, our survival now no longer depends on our ability to adapt to new scenarios. We must enhance the ability to create and to exercise influence in these new scenarios. It is more important to be the change than to adapt to change.

1.2 Uncertainty

There are several trends, but a lot of them are "*game changers*" (factors that could drive trends in a positive or a negative way, but we do not know the direction) that make very difficult to know the future evolution. Many of the changes that will occur, very frequently, will be disruptive, showing that the past is not necessarily a predictor of the future, and hindering our ability to be prepared for coming scenarios. While uncertainty makes forecasting less reliable, the decisions still have to be made. "*The future, like everything else, is no longer quite what it used to be*", Paul Valery said in 1937.

Uncertainty is a key characteristic of our societies, generating fears that must be managed by governments and security institutions.

1.3 Complexity

Too many factors are involved in current phenomena; some of them are both causes and effects at the same time. Everything is connected. We face a complex set of threats emanating from state and non-state actors. Each event is conditioned by a multiplicity of quantitative and qualitative causes and factors inter-related to third events. Interdependency, diversity and difficulties to understand and mitigate the problems are arising facts.

1.4 Ambiguity

Casual relations are not clear. It is difficult to determine the combination of factors or drivers that will drive the future in a positive or negative direction. It is very difficult to find correct answers to classical questions: what, why, who, when, where and how. Misinterpretations and plurality of possible meanings are cause and effect of confusion in the current decision-making environments.

Taking into account these characteristics and the new crime scenarios that should be faced, it is possible to state that organised crime is clearly a context of rapid change, greater complexity and genuine uncertainties. A review of some of the new events that have been taking place all around the world shows a great evolution of the threat and the continuous emergence of new battlefields (Fig. 1).

One of the implications of addressing these situations is that their complexity also determines the complexity of the problems to tackle. Each challenge becomes unique and new compared to previous ones (Grint 2010), challenging the responsiveness of institutions. The possible solution must be adopted as quickly as possible, with the difficulty of dealing with problems not faced previously. The need to identify and evaluate the success of the possible decisions to adopt, but also the unintended consequences, becomes indispensable. If this is not done, the decision-making can lead to additional problems that enhance what we are trying to combat.

In this sense, the difference between old and new problems is not that the former could not have a complex solution, but that they could be solved under the application of techniques and processes already known, while new problems have a lack of optimal known solution. In addition to its nature and composition, the new problems can never be tackled without knowing the concrete context in which they develop.

A way to analyse these scenarios is trying to know how many alternative interpretations can include their meaning and how many approaches are possible to identify entities, stakeholders, indicators, drivers, enablers, enhancers, vulnerabilities or motivations. This should be the first stage to establish priorities, identify constraints and define interests from own institutions. This exercise requires the application of non-formal analysis techniques (as is discussed in Chapter "Organised Crime, Wild Cards and Dystopias" of this book) that allow to understand, in a rigorous and fluid way, the possible changing landscape in which we will need to detect, dismantle and deter the future organised crime.

Several of the most traditional elements in the performance of the organised crime are changing, and several of them are not essential in these moments (Fig. 2):

The starting point to deal with these challenges is the identification of a set of trends and megatrends that run during the current situation. As this criminal phenomenon has an international character, this identification should be global, holistic and based on a comprehensive approach.

Following the extensive use of VUCA has emerged another acronym as an antonym significance trying to focus on the perspective from which these

V E L O C I T Y	- Digitalization, connectivity, trade liberalization, global competition, business model innovation - Highly adaptable to supply and demand factors, operating environment. Ability to diversify business adapting to economic drivers and government countermeasures. - Quickly targets mutations (Makareno, 2004).
U N C E R T A I N T Y	- Expansion of virtual criminal undergrounds in any place without physical presence. Cybercrime. - On-going improvements to use anonymous protocols, password authentication techniques, and encryption techniques. - New ways to finance activities. Anonymous contributions across Internet, sophisticated frauds, virtual currencies (Europol, 2015). - Non old-style hierarchies (Rand Corporation, 2001).
C O M P L E X I T Y	- A broad set of competitors and customers, extended supply chains, segmented markets, disruptive technologies. - Intersection of different types of organizations, increasing convergence of terrorism and crime, as service suppliers, sources of new incomes, or simply partners on training and sharing targets, among others. Growing diffused ties (Chatham House, 2015). - More resources (economic, weapons, human resources) than many public security forces. - Crime as a service. Networks offering knowledge and resources to small criminal groups or individual.
A M B I G U I T Y	- Social media and big data are used to monitor, identify and neutralize competitors, sell products, recruit, but also to manage public relations amplifying its public profile (World Economic Forum, 2015). - Ability to operate as a large multinational corporation. A way to establish a "criminal state" model. - More targets. Increasingly social acceptance in weakened societies where new services can be provided (Europol, 2015). - While the previous scenario takes place its capacity to become a new actor gives more power to organised crime groups. There are more spheres of influence using corruption to maximise profits. - Organised crime can act as an agent in industries that depend on natural resources (Europol, 2015).

Fig. 1 Some VUCA indicators on current serious transnational organised crime business (Makareno 2004). Jéssica Cohen 2016

environments must be understood, "VUCA Prime" (first Introduced by Bob Johansen): vision, understanding, clarity and agility. It is configured as a set of inexorable skills needed in the present and future times of our societies (Fig. 3).

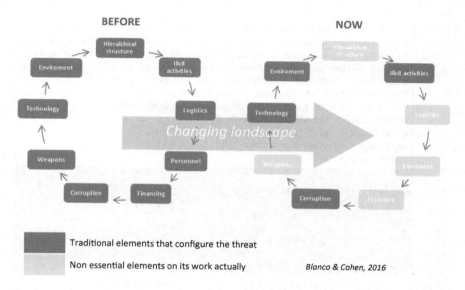

Fig. 2 Traditional versus current elements that configure the threat. Jéssica Cohen 2015–2016

2 A Methodological Approach. A Criminal Intelligence Process

Before defining new policies, an analysis of the current situation of the phenomenon and its possible evolution is needed. Critical thinking, loads of imagination, creative foresight and horizon scanning methodologies would be the pillars of the research.

Policies are usually led by events and by social perception of risk. We propose a holistic approach that integrates the lessons learned from the past with modern foresight methodologies, intelligence analysis, evidence-based policing and decision-making models. It is possible to manage our uncertainties in the present, a time and a matter in which perhaps we feel lost, but we must be sure that we are walking in a correct direction.

This section offers a methodological model for analysing the phenomenon of organised crime. There is a need to understand the environment and its evolution, which determines facilitators, enhancers and inhibitors in the present and the future of the criminal phenomenon. The proposed model combines EUROPOL methodology used in the SOCTA reports with an intelligence process adapted to our intelligence information society. New technologies have changed the contents of the traditional intelligence cycle.

In December 2015, the Council of the European Union published a document titled "Serious and Organised Crime Threat Assessment 2017. Revised methodology", which was validated by the EU Standing Committee on Internal Security (COSI) on 24 November 2015. SOCTA is a strategic report assessing and prioritising SOC threats and risks in the EU, which contains an analysis of crime areas

+ Ability to predict the effects of a situation **−**	**CLARITY over COMPLEXITY** Even chaos can make sense. Generate knowledge maps. Make a dynamic tracking of the existing analyses to detect new evidences (monitoring). Understand each phenomenon from within and from the global perspective simultaneously. Do not use simplistic, mono-causal or mere chance explanations, trying to answer all possible questions.	**VISION over VOLATILITY** Think in future as a habit. Imagine scenarios and analyse them in a back-casting process to detect indicators, in order to avoid future risks and threats. The objective and methodology applied must be clearly defined.
	Adaptive thinking Lateral thinking General / holistic approaches Knowledge Management Observation Curiosity Starbusting creativity techniques	Learn to learn Knowing how to unlearn Continuous training Antifragility (N. Taleb, 2013) Creativity Agility Cognitive adaptability
	AGILITY over AMBIGUITY Maximize the ability to learn, make mistakes, communicate, respond and adapt. It requires rapid problem solving and constant decision-making. It must be proactive and be focused on the problem to anticipate the effects even before adopting the answer.	**UNDERSTANDING over UNCERTAINTY** The phenomena of organised crime we face must be fully understood. The answer should go beyond our own previous experience and knowledge. It needs to build knowledge networks, with trust and credibility, and use new technologies to strengthen the whole process.
	Critical thinking Experimentation Learned lessons Learn to doubt Dismisses the superfluous Self-driven learning Finding solutions Proactivity	Transparency Confidence Intelligence of the crowds Collaboration / teamwork Creating scenarios / simulations Idea Generation Resolution / decision-making Validation of acquired knowledge

− Knowledge about a situation +

Skills and tools, individual and organizational, to work in chaos.

Jessica Cohen, 2015-2016

Fig. 3 Skills in VUCA times (Taleb 2013). Jéssica Cohen 2015–2016. *Note* The − and + signs in the two axes, 'Ability to predict the effects of a situation' and 'Knowledge about a situation', indicate the degree in which we can measure, respectively, the ability and the knowledge, ranging from less (−) to more (+)

(criminal activities), organised crime groups, and the environment of organised crime. EUROPOL defines "environment" in a broad way, with three sub-aspects: enabling factors (vulnerabilities and opportunities for criminals), the geographical scope and the effects on society. These factors are called crime-relevant factor (CRF). For these purposes, it implements a horizon scanning methodology, a set of indicators, and a focus on a PESTLE model.

We adopt the methodology proposed by Blanco and Cohen (2014), analysing the future of terrorism. This model is used by the Centre of Analysis and Foresight of Guardia Civil, a Spanish Law Enforcement Agency with military nature, to analyse the future of different criminal activities and groups and other threats to national security: violent radicalisation, irregular immigration, the evolution of international conflicts or social polarisation.

Although this model offers a strategic approach, it provides the building blocks for the construction of an early warning system, which are one of the possible ways to link strategic and operational analysis.

The model consists of five phases, which in any case are not cyclical or successive but overlapping and generating continuously feedback: planning, monitoring, analysis, visualisation and evaluation.

2.1 Planning

Planning, an obligation in any project, requires determining the object of study, identifying information needs, selecting initial sources, allocating resources and setting deadlines.

2.2 Monitoring

By names such as monitoring or scanning, we fundamentally define continuous information collection systems. New technologies allow the automation of classical tasks. Collecting information, classification, evaluation and integration are supported by the use of semantic search engines, crawlers, content curation applications, entities extractors, translators, converters of voice into text, treatment of unstructured information (in heterogeneous formats), and connecting internal databases with open-source data and cloud computing systems. All these tasks can currently be performed at once, while the information is collected (Fig. 4).

2.3 Analysis

For the purposes of this model, three levels of analysis are distinguished. First is the classical analysis as has been reflected in studies and professional development of

MONITORING	OBJECTIVES	RESOURCES
Requirements	• Organised Crime Areas • Organised Crime Groups • Organised Crime Environment (facilitators, geography, effects)	• Literature review • Previous information and data • Previous risk analysis • Detected trends
Sources selection	• Data • Information • Previous "intelligent" knowledge • Experience • Imagination and creativity	• Literature review • Sources management: opens sources, human sources and technical sources
Collecting and managing information	• Automatisation of collection and management of information • Evaluation of information • Integration of information	• *Environmental Scanning* • *Crawlers* • *Content curation* • Evaluation: reliability of the source • Evaluation: credibility of the information • Entity Extraction (dates, places, companies, persons…) • Relationship between entities • Network analysis • Data bases and document repositories
Classification of information and knowledge management	• Selection and classification of information • Knowledge management	• PESTLE • Timelines • Mind maps • Semantic search and semantic analysis

Fig. 4 Monitoring information (Cohen and Blanco)

intelligence, which uses both social science methodologies and structured analysis techniques (Heuer and Pherson 2014). Second is the predictive analysis based on technologies that enable the management of large amounts of data. And third is, and not less important, although possibly less used and less valued, the foresight analysis. Several precautions are taken against this kind of analysis because of its longer term and its greater degree in regard to uncertainty (Fig. 5).

In the following lines, we will detail the methodologies that lead, from our point of view, to a better understanding of this criminal phenomenon and its possible evolution.

Macro-environmental Factors Driving Organised Crime

ANALYSIS	OBJECTIVES	RESOURCES
Descriptive Analysis	• Evaluation of information: • Integration • Analysis	• SWOT • Lessons learned • Analysis of Competitive Hypothesis • Analysis of variables • Analysis of actors • Social Network Analysis • Statistical Analysis • Structured Analytical Techniques • Analysis of Crossed Impacts • Environmental Scanning • PESTLE
Predictive analysis	• Prediction of future magnitudes of OC: when, where, who	• Correlation • Regression • Multivariable Analysis • Big Data • Artificial Intelligence • Geo-localisation • Predictive Policing • Criminal profiling • Geographical analysis
Foresight analysis	• Trends detection • Identification of indicators • Disruptive analysis • Alert system • Scenarios planning	• Trend analysis • Trend evaluation • PESTLE • Delphi • Indicators • Creativity Techniques • *Wild cards* • *What if* • Early Warning System • Morphological analysis • Scenarios

Fig. 5 Analysing (Cohen and Blanco)

(a) Environmental Scanning

We understand environmental scanning as a systematic and formal process with two main objectives:

- Understanding the nature of change in the environment
- Identifying opportunities, challenges and future developments

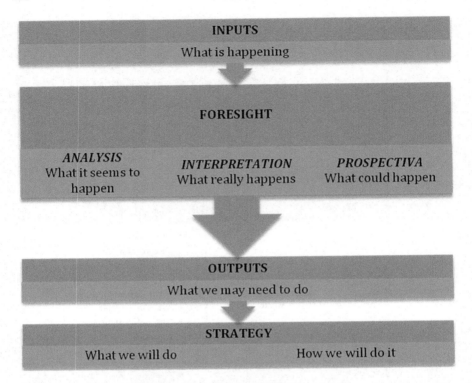

Fig. 6 Environmental scanning structure (Cohen and Blanco)

Its main objective is to answer the questions raised (what is happening and what could happen), determining what we know and what we do not know about the phenomenon studied, with strategic purpose (Fig. 6).

(b) PESTLE

PESTLE has traditionally been a part of environmental scanning and monitoring activities in most organisations. It is a process of analysis that aims to study the political, social, economic, technological, legal and environmental issues affecting a sector or a field, configured as a first step in many future studies (Bensoussan and Fleisher 2013). This method is sometimes called STEEP. It is a way to break the general environment into subcategories or segments, aiding to follow an analytic process in which we can research each of the components of a phenomenon with strategic management purposes.

The process involves understanding each of the mentioned segments, and how they affect the object of the study, answering the following questions: What are the current key events and trends in this segment? What evidence supports the existence of these trends? How have these trends evolved? What are the nature and degree of change on turbulence within trends? How do the trends affect organised crime?

PESTLE represents a systems-of-systems analysis approach (SoSA), also known as a "Federation of Systems" (Svendsen 2015a, b).

(c) Semantic Analysis

Semantic search seeks to improve search accuracy by understanding the searcher's intent and the contextual meaning of terms as they appear in the searchable dataspace (series of databases), whether on the web or within a closed system, to generate more relevant results. Semantic search systems consider various points including context of search, location, intent, variation of words, synonyms, generalised and specialied queries, concepts matching and natural language queries to provide relevant search results (Tony, J, March 15, 2012 "What is Semantic Search"). Semantic searches allow to classify directly the collected information in order to match entities from other pieces of information. It uses several methodologies: concept mapping, graph patterns, logics and fuzzy relations and fuzzy logics.

(d) Trend Analysis

This method aims to identify trends and to detail the forces that are influencing, in what direction and at what speed (i.e. the intensity of change). A trend is a pattern of observable change, a set of processes that are not easily changed and that will continue in the future. Trends are affected by drivers. A driver is an agent or factor that guides a change. From each driver, it is possible to identify indicators that help measure and assess its impact. This fosters a model to follow up on a matter that is also complemented with a determination of the effects on the phenomenon researches, their impact and probability of occurrence. The paper by Blanco and Cohen (2014) on the future of the fight against terrorism in Europe is an example that is based on the model chosen by RAND Europe (2013) for the analysis of the future of society in Europe in 2030 and on Lia's model (2005) to study the future of terrorism, as well as the models used by the Proteus programme of the American intelligence services.

Remarkable is the use of indicators in this methodology, which allows the establishment of a model of continuous monitoring and evaluation.

(e) Indicators

An indicator is a measurable variable used as a representation of an associated (but non-measured or non-measurable) factor or quantity.

An indicator must be specific (precise and unambiguous, so it is clear what it is that you are aiming to achieve), measurable (there should be a clear and transparent measure of success), relevant (it should reflect what the organisation is trying to achieve), timed (it should be clear when the target should be delivered), appropriate (to the subject at hand), economic (it should be available at a reasonable cost), adequate (it should provide a sufficient basis to assess performance) and monitorable (it should be amenable to independent validation).

For further insights into environmental scanning, PESTLE and indicators, see Chapter "Foresight and the Future of Crime:Advancing Environmental Scanning Approaches".

2.4 Visualisation

The dissemination of intelligence products remains an unstoppable trend towards graphics and intuitive ways, to the detriment of classic and extensive textual reports. Screens that can display and analyse the evolution of indicators, warning systems to alert of changes in patterns or upon detection of a "weak signal" (Chapter "Foresight and the Future of Crime:Advancing Environmental Scanning Approaches" for more information), infographics, the possibility of specific statistical analysis, specific searches and advanced queries developed by the analyst or decision-maker, or the simulation of future scenarios changing some of the variables are part of the possibilities that new technologies enable and that facilitate the task of analysts and decision-makers in an environment of information overload (infoxication) and shortage of time.

2.5 Evaluation

A pending task is the assessment of environmental analysis and future studies and research. At least, a basic analysis could be carried out based on:

- The model "Analytical Quality Framework" of the Defense Intelligence Agency (DIA), and the adaptations of Boardman and Pherson applied to intelligence products, and Blanco and Cohen (2014) for counter-terrorism research.
- Clark (2013), based on intelligence, proposes the consideration of the following criteria:
    - Credibility. The result makes sense.
    - Relevance. The process answers the questions.
    - Urgency. Need to establish deadlines for action.
    - Usability. Possibility of use by the decision-maker
    - Technical quality. Existence of a methodology.

3 The Future Macro-Environment of Organised Crime. Trend Analysis

The recent report from Europol (2015) on *Exploring tomorrow's organised crime* characterises the *key drivers for change and their impact on serious and organised crime* as follows: "Organised crime will undergo profound and significant changes over the next decade in response to the availability of new technologies, changes in the environment such as economic challenges or developments in society and in response to law enforcement actions. Organised crime will undergo these changes whether or not experts agree on a new definition of organised crime and it is imperative for law enforcement to seriously consider the factors and driving forces that will shape serious and organised crime over the coming years". EUROPOL detects eight great trends of change:

- **Transportation and logistics**

Innovation in transportation and logistics will enable OCGs to increasingly commit crime anonymously over the internet, anywhere and anytime without being physically present.

- **Data as a commodity**

The increasing exploitation of big data and personal data will enable OCGs to carry out complex and sophisticated identity frauds on previously unprecedented levels.

- **Nanotechnology and robotics**

Nanotechnology and robotics will open up new markets for organised crime and deliver new tools for sophisticated criminal schemes.

- **E-waste**

E-waste (i.e. discarded electronic or electrical devices or their parts, also called electronic waste) Without the necessary legislative and law enforcement responses, the illicit trade in e-waste is set to grow dramatically in the near future in terms of both quantities traded and the quality of the methods used by criminal actors engaging in this activity.

- **Economic disparity within the EU**

Economic disparity across Europe is making organised crime more socially acceptable as OCGs will increasingly infiltrate economically weakened communities to portray themselves as providers of work and services.

- **Increased competition for natural resources**

OCGs will increasingly attempt to infiltrate industries depending on natural resources to act as brokers or agents in the trade with natural resources.

- **The proliferation of virtual currencies**

Virtual currencies increasingly enable individuals to act as freelance criminal entrepreneurs operating on a crime-as-a-service business model without the need for a sophisticated criminal infrastructure to receive and launder money.

- **Demographic change in the EU**

OCGs will increasingly target as well as provide illicit services and goods to a growing population of elderly people, thus exploiting new markets and opportunities.

We have applied the analytical framework used by RAND Europe in its research "Europe's Societal Challenges. An analysis of global societal trends to 2030 and their impact on the EU" that is consistent with the Development, Concepts and Doctrine Centre (DCDC) report of the UK Ministry of Defence (2014), titled "Global Strategic Trends out to 2045".

First of all, we selected the trends observed through a literature review process. We identified documents about the future of our societies and the future of crime. We selected political, economic, social, technological and environmental trends, based on the following sources:

- "Europe's Societal Challenges". RAND Europe
- "Global Strategic Trends out to 2045". DCDC
- "Global Risks 2013", "Global Risks 2014", "Global Risks 2015" and "Global Risks 2016". World Economic Forum.
- "Global Trends 2030". National Intelligence Council.
- Internal documents of the CAP, Guardia Civil, based on a general literature review about the future, and the selection of the main trends about the future with a PESTLE/ SWOTmodel.

Using the same documents and helped by experts meetings in the headquarters of the Centre of Analysis and Foresight (CAP) of the Guardia Civil (Spanish Law Enforcement Agency), we identified a set of drivers (factors that influence or causes change), indicators and those outcomes that could be related to terrorism, that could act as causes or factors, as well as produce threats and risks.

As mentioned above, a trend is a set of processes that cannot be easily changed, a discernible pattern of change. Trends were born, have been pushed or depend on the present and will continue in the future. A driver is an agent or factor, which drives a change forward, and the indicators are the different variables that explain it.

In our process, we partially follow the model of Lia (2005). Although he uses a PESTLE model for the study of the future of terrorism, his framework is useful as a general approach for other phenomena. The main advantage of this proposal is that it defines a framework to analyse the environment regarding the potential sociopolitical changes enabling the evolution of terrorism.

Lia basically mentions that there are factors, such as international relations (leadership, proliferation of weapons of mass destruction, democratisation, fragile states, multilateralism, peace support interventions, non-governmental actions), economic factors (inequality, relationship between economy and politics, organised crime, energy), demographic factors (growth and migrations), ideologies and technologies, that would allow identifying the causes of terrorism and predicting the future (target patterns, terrorism level, deadliness, ideological motivations, geographical location, etc.). A similar model is the one we will use in this chapter to analyse the future of organised crime providing a schematic view on trends and security outcomes for different PESTLE dimensions, namely political, economic, social, technological and environmental trends. For each trend, the following will be shown: drivers, indicators, degree of evidence (***: high, **: medium, *: low), time range (S: short, M: medium; L: long) and outcomes which were qualified by degree of uncertainty (H: high; M: medium; L: low).

Those risks with low probability (because of the high competencies or resources needed to plan and execute an attack or heist, historical trends or incidents that go beyond our imagination) and high impact (because they generate high costs, terror, casualties or dangers) are what we call "wild cards". Different from those called as "known unknowns", the things that we know will happen although we do not know exactly when or what effects will result of them.

There is a key factor that could act as a "game changer": counter organised crime policies, that, as we argue, it is not possible to determine whether they act as a pull, a push or a neutral factor because the EU does not follow an evidence-based policing system.

Political trends

Environmental disorder, instability in the exercise of power, fragile states and the existence of armed conflicts without end are a framework that generates enormous opportunities for organised crime groups. Situations that are exploited for profit, as it is happening with the humanitarian and refugee crisis. These contexts additionally offer protection to criminals operating in geographical areas where the weak institutions, especially in security, defence and justice, limit their persecution.

Naím (2013) argues that power is not merely shifting and dispersing. It is also decaying. Institutions and organisations in power today are more constrained in what they can do with it. Continuous leaks are showing that transparency is a key factor of change that should lead to a new way of executing power. The existence of micropowers can "topple tyrants, dislodge monopolies and open remarkable new opportunities, but it can also lead to chaos and paralysis". Non-governmental actors hold a lot of power: multinational companies, lobbies, insurgencies, criminal organisations, etc.

The transnational dimension of organised crime is favoured by globalisation, in all its orders. International organisations have no answers to these threats, while states face serious limitations to fight against transnational risks.

Corruption is a generalised phenomenon, with local, national and international implications. "In every environment in which corruption operates, it is central to the rise and perpetuation of crime, terrorism and other social ills", states Shelley (2014). The reverse is also true. Organised crime produces several effects, rising instability and corruption, in a continuous cycle with causes and effects. The enabling environment of terrorism for crime and terrorism can be found all over the world (Fig. 7).

TRENDS	DRIVERS	INDICATORS	E	TIME	OUTCOMES	U
Crisis of power	Obama doctrine: "Leading from behind"; crisis of the power of nations; new international balance of power; multi-polar world and diffused power; post modern times (Bauman); lack of EU international leadership; world processes of democratisation.	Global Peace Index (GPI); Fragile States Index (FSI); inequality gaps between countries; power indicators (GDP, military power, etc.)	**	S/M/L	Global governance failure; anarchy; failed and weak states; corruption; organized crime; terrorism; long and permanent low level conflicts out of control	H
Growing power of non state actors	Terrorism and organized crime groups, individual power, multinationals, "markets", cities, NGO.	Number and spread of organizations. Number of International conflicts (economic, insurgencies, financial) between states and non state organizations. Global Peace Index	***	S/M/L	Conflicts, organized crime, insurgencies, corruption, terrorism. Social unrest (states can't respond to the demands of its citizens)	
Growing local and regional conflicts	EU surrounded by unstable regions (Middle East, North of Africa); military participation of EU countries; peacekeeping operations; War on Terrorism.	GPI; FSI; presence of European countries troops in conflicts	**	S/M/L	Fragile states; corruption; organized crime; terrorism; intrastate conflicts; insurgencies.	H
State fragility	Several states are in a fragile situation, can be taken by criminal actors, insurgencies or terrorists (e.g. Libya). Not only important the conflict, but the post-conflict management.	Fragile States Index	***	S/M	Corruption, insurgencies, terrorism, social unrest, interstate and intrastate conflicts, organized crime, illicit trafficking	H

Fig. 7 Political trends (Cohen and Blanco)

Crisis of democracy	Institutional credibility in crisis; failed democratisation process; demands of transparency; new desires about political participation; activism and hacktivism	Surveys (institutional credibility); Transparency Index, Open Data Index; legitimization of the state index	**	S/M	Social unrest, lack of stability, reduction of legitimization of the state	M
Corruption	Crisis of democracy, conflicts, state fragility, crisis of state power, proliferation of non state actors, lack of controls, impunity	Corruption Perception Index	***	S/M	Social unrest, lack of stability, reduction of the legitimization of the state, attractiveness towards illegal traffics.	H
Growing surveillance	Security concerns about privacy and cyberspace; legal systems of control	E-government and open data indexes; sentiment analysis output; proliferation of leaks	**	S/M	Multi surveillance from the states and from citizens; leaks; social unrest; people trying to hide	M
Growing popularity of grassroots and populist movements, nationalism, extremism	Decline trust in institutions; Diverging global attitudes; anti globalization movements; nationalisms; Euro-scepticism ascendant; historical factors	EU elections results; national election results; number of terrorist attacks or incidents; confidence in institutions surveys.	***	S/M	Social unrest; Political violence; Terrorism; OC for financing activities	H

Fig. 7 (continued)

Economic trends

Organised crime has a clear objective: economic profit. All the conditions that determine the economic and financial situation generate new opportunities. Organised crime finds facilitators in the economic environment: new services, new resources, new routes and transports, new payment methods.

Criminal groups have huge resources, both personal and economic. They have the ability to hire the best economists, financiers, money launderers or lawyers trying to take advantage of all the gaps in tax and legal systems, and the existence of tax havens.

Some economic factors generate greater chance of uptake, being the black and criminal market an incentive for young people without other opportunities.

Modern criminals are innovating "not just technologically but in their business models as well" (Goodman 2015). They know the latest corporate strategies, supply chain management, global logistics, creative financing and consumer needs. Crime Inc. is a global enterprise, a full-service and a multiproduct company.

Crime as a service (CaaS) is a new business model, with the objective of keeping the crime factory humming and bolstering the underground economy, offering, for example prepackaged tools for phishing, spam, fraud, data theft and other crimes (Fig. 8).

Social trends

Demographic trends, changing values in our societies, empowerment of citizens, urban development, work and leisure-related mobility, migrations and access to knowledge are some of the variables that make up the environment.

The social change that we are witnessing cannot be understood apart from the mutual influences with political, economic and technological variables.

Citizens express new demands, new needs, new expectations, within the framework of a new way of understanding the world and acting in it (Fig. 9).

Technological trends

Whatever the technical innovation, criminals are quick to adapt to them (Goodman 2015). They use information and communication technologies to collect information about their targets. It is clear that they develop early warning systems to notify them news that could affect their business, and software to track clients, providers and targets. Advanced technologies are used for money laundering purposes.

Criminal groups use applications to protect their identity and to cover his tracks. Tor could be only one of these examples. The use of crypto currencies is a way to hide activities and payments.

In a world where everything is hackable, the Internet of Things offers great new opportunities to these criminal groups. Chips and sensors connected to internet could be a way to access to personal data. All physical objects will be assigned an IP address. But not only that, they can also be pathways for cyber attacks on our way of future life (smart cities), infrastructure (electrical systems), against technologies (autonomous cars) or against persons (heart valve).

But there are thousands of new criminal uses of new technologies. Three-dimensional printers for fraud or firearms, and robots could do criminal activities with a lower risk of exposure to criminals, drones are being used by cartels (drug trafficking) and terrorist groups (Hezbollah), artificial intelligence brings us closer to some futuristic dystopias that we see in the movies. Mental hacking is an emerging risk. Biotechnology proposes the creation of bio-cartels and multitudinous new drugs production and consumption, including those that enhance human skills. Synthetic biology holds the potential to disrupt the way they cartels do business. Taking the genetic code of classical drugs avoids the need to cultivate plants (Fig. 10).

TRENDS	DRIVERS	INDICATORS	E	TIME	OUTCOMES	U
Growing unemployment	Shortfall or slow employment creation; labour exploitation	Unemployment rate; youth unemployment rate	**	S/M	Social unrest; inequality; populism and extremism; radicalization; guettos and minorities integration; resurgence of leftist anti globalisation ideologies	L
Inequality inside states and between rich and poor countries	Disparities inside EU; Unemployment; development; education; health	Income indexes; Gini index; income per capita; Human development index	**	S/M		L
Economic and financial crisis	Lack of opportunities; reducing social benefits; unemployment; inequality	Gross Domestic Product (GDP); social budget evolution	***	S/M		H
Decreasing global poverty	Developing countries	GDP; income per capita; Human Poverty Index; Multidimensional Poverty Index	**	S/M/L	More opportunities	M
Increasing spread and power of financial markets, mobility of capitals, humans and investments	Globalization. Transactions on line. Tax havens. Lack of controls	Volumes of transactions	***	S/M/L	Money laundering, corruption, fraud, tax evasion	H
Proliferation of virtual currencies	Spread of virtual currencies, like "bitcoin". New financial markets. Controls and restrictions	Prices, transactions, number and volume of new currencies	**	M/L	Money laundering; lack of control over financial operations; fraud; robbery; tax evasion	M
Tax havens	Proliferation of tax havens, perceived as a way to hide incomes from illicit activities or tax evasion or avoidance	Number of tax havens; Controls exercised; Number or scandals known; quantity of funds in tax havens	***	S/M	Social unrest, inequality, lack of state power, money laundering	H
New commercial routes, with new transports	New desires, new products. New international treaties: EU-USA, USA-Asia	Volume of transactions between countries; evolution of levels of violence, drugs use, homicides.	***	S/M/L	Organized crime, illicit trafficking, social unrest	M

Fig. 8 Economic trends (Cohen and Blanco)

TRENDS	DRIVERS	INDICATORS	E	TIME	OUTCOMES	U
Global population growth	Growing life expectancy; growing fertility in developing countries	Fertility rates; life expectancy	■■■■	M/L	Stain on natural resources and food; migration floods	L
Population ageing in high and middle income countries and family changes	Growing life expectancy; high fertility rates; less infectious diseases; elderly citizens; increasing in one single person ways of life; new forms of cohabitation; youth bulges on parts of the world	Fertility rates; life expectancy; % of single parent households; divorce rates; average households size; poverty by household; old age dependency ratio, health care costs; proportion of young population	***	S/M/L	Risks of poverty and social exclusion; migration to EU; social unrest; pressure for democratic reforms; lack of opportunities; radicalization.	M
Decline working population in EU	Fertility decreases; longevity increases	Fertility rates; population growth; dependency ratios; % older people	***	S/M/L	Migration to EU; pressure for democratic reforms; decreasing political EU influence; social unrest	L
Migrations	Diversity in migration flows; attractiveness for migrants: economic and employment opportunities; social and family networks; rights and liberties; attitudes to migrants; residential distribution of migrants; conflicts; second and third generations integration; "diaspora" communities	Global Peace Index; Human Development Index; Racism surveys; migration rates; integration indicators; education level and success rate; migrant unemployment rates; diasporas communities evolution	***	S/M/L	Cultural polarization; problems of integrations; ethnic and religious conflicts; stress to social welfare; racism; skill gaps for job access; guettos; lack of opportunities; extremisms and radicalization; hate crimes; inter communal violence; home-grown terrorism	H

Fig. 9 Social trends (Cohen and Blanco)

Global urbanization	Economic opportunities; globalization of travel and transports and mobility	Urban population rate	***	S/M/L	Social exclusion; guettos; radicalization; resource scarcity; urban warfare	M
Human development	Better education, health and wealth	Human Development Index Education failure	***	M/L	High variations in EU; opportunities that can be applied or failed	L
Social conflicts	Marginalization, urban guettos, polarization of the population, cultural fragmentation, extremism and radicalization, crisis of democracy, economic crisis, lack of credibility on institutions	Global Peace Index. Number of demonstrations, strikes. Spread of social movements	**	S/M	Social unrest, insurgencies, terrorism, hate violence, xenophobia, racism	L
Individual empowerment	Internet; social media, networks,; education on line; information and knowledge society; new ways of organization avoiding intermediation (hotels, taxis, books, 3D printing)	Internet and social media use; evolution market of crowd services and 3D printing	***	S/M	New opportunities, but organized crime an terrorist opportunities too in communication and recruitment: loner terrorism, foreign fighters; a lot of business affected; unemployment; Intellectual property piracy	H

Fig. 9 (continued)

Environmental trends

The environment appears as one of the greatest risks in all current reports regarding the future of the world. Our societies raise awareness about its importance. The scarcity of resources and its geostrategic character makes this sector a focus of attention of organised crime groups. Any demand for a product or resource above its offer, or the possibility of offering it a lower cost, involves new criminal opportunities (Fig. 11).

Although there is not a clear category in the PESTLE analysis to consider other factors, we think that talking about organised crime we need to take in consideration geographical, historical and criminal factors.

TRENDS	DRIVERS	INDICATORS	E	TIME	OUTCOMES	U
Development of information and communication technologies	Internet and social media; internet of things; semantic web; cloud computing	Technological foresight; analysis of impact of new technologies in security	***	S/M/L	Growing cyber-threats: cyber-terrorism, cyber-attacks, cyber-crime; critical infrastructures risks; espionage; intellectual property attacks; privacy; technological inequality	H
Big Data and predictive systems	Internet and social media; internet of things; semantic web; cloud computing	Volume of information (internet, social media), variety, and velocity	***	S/M/L	New opportunities to understand and study matters like terrorism. Privacy: terrorism and effects on liberties	H
Technological development: means of transportation; nanotechnology; 3D printers, cyborgs, Google glass, CCTV...	Access to technologies; economic cost; drones	Technological foresight; analysis of impact of new technologies in security; study of cases and news	***	S/M/L	Push and pull factors for criminal and terrorist use and for a counter-terrorism use. Dual use civil and military and possibility of access by terrorist groups. Gap between legislation and technology. Privacy	H
Transportation and logistics	Mobilisation (people, resources). New means of transport, new routes.	Evolution of the sector	***	S/M/L	Organized crime, trafficking, corruption	H
Smart cities	Technologies; social and political change and trends; internet of things	Smart cities index	***	S/M/L	Cyber-threats; critical infrastructures targets	H
Nanotechnology and robotics		Evolution of the markets; collection of problems caused	**	M/L	Conflicts, privacy, ethical and legal questions, new crimes	M

Fig. 10 Technological trends (Cohen and Blanco)

TRENDS	DRIVERS	INDICATORS	E	TIME	OUTCOMES	U
Growing resources demand	Cultivation; new technologies; economic and population growth; urbanization degree; food, water and energy scarcity, e-waste	Prices of resources (oil, minerals, food); international conflicts	**	M/L	Conflicts; resources scarcity and competition; food crisis; humanitarian crisis; migrations	H
Climate change	Determining the direct effect of climate change in natural disasters and other effects	Climate effects: droughts, floods, tsunamis,...	**	S/M/L		H
Pandemics	Urbanisation; globalisation; travels	Evolution of incidents	*	S/M/L		H
Natural catastrophes	Floods, Volcanoes, earthquakes, tsunamis	Evolution of incidents	*	S/M/L		H
Attractiveness sector for OC groups	New products, new markets	OC indicators. Seizures. Police operations. Arrested	**	S/M/L	Trafficking. Corruption. Insurgencies. Money Laundering	H

Fig. 11 Environmental trends (Cohen and Blanco)

- Geography is a key driver for organised crime: long borders, extension of coasts, high mountains, jungles or deserts facilitate the activities and movements of OCGs.
- Historical factors: impunity, tradition of organised crime groups, historical trafficking routes are key factors in the present and future of the phenomenon.
- Of course, there are specific criminal factors, indicators and information that we should monitor, as a way to identify trend patterns and possible changes: links between OC and terrorism, police efficiency, judicial system, security perception, prisons system, weapons, extradition policies, criminal groups, black markets, illicit traffics, corruption, logistics for organised crime, money laundering, black economy, extortion, kidnappings and related crimes.

4 Case Study: Cocaine Trafficking

Cocaine trafficking provides an excellent basis to create a story-based narrative which explores each stage of the criminal journey of the product or person revealing signs and indicators from a number of different sources.

Cocaine trafficking is one of the EU's priorities for the fight against serious and organised crime between 2014 and 2017 (Council of the European Union, Brussels, 28 May 2013). The SOCTA 2013 provided the basis on which council agreed nine SOC priorities for 2013–2017. One of them is cocaine and heroin, with the aim to reduce trafficking to the EU and to disrupt the OCGs facilitating the distribution in the EU (Europol 2013).

The aim of this use case is to build a framework for the identification, monitoring and evaluation of political, economic, social, technological, environmental and legal trends.

In order to establish a PESTLE model of analysis, the first stage was to point out the general factors involved. The identification of key factors followed a methodological approach: literature review, expert opinion, panels and workshops. These factors could be classified following a trend analysis model: trend, drivers and indicators.

Two ways of managing sources to identify other trends were proposed:

- Semantic search in internet. The objective is not only to find information but also to interpret it in a way that could facilitate the work of security analysts.
- The establishment of an indicators system.

4.1 Semantic Search

With the technological opportunity to collect, classify and interpret information in internet, the Guardia Civil and the University of Aalborg selected a representative case from the news. They analysed several pieces of information that could content several of the key elements about cocaine trafficking: origin, destiny, location of the seizure, criminal group, concrete characteristics, kind of drug, weight or dates, showing that an intelligent entity extraction system can help to analyse this phenomenon (Fig. 12).

The analysis of this piece of information allowed to "translate" the meanings that a semantic system should search in the existing contents in internet. In this case (Fig. 13):

This system and the information available in the web would allow to collect, classify and detect changes in patterns and to match data, for example if in other parts of the world a package of cocaine marked with "00" appeared.

Macro-environmental Factors Driving Organised Crime

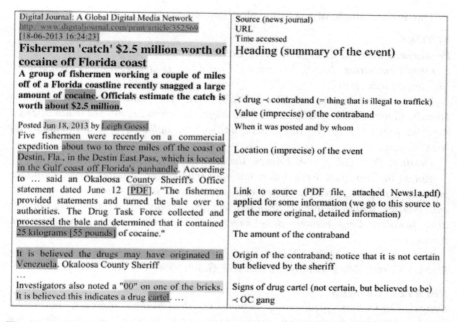

Fig. 12 A selected piece of information about cocaine trafficking (Henrik Legind Larsen, Aalborg University)

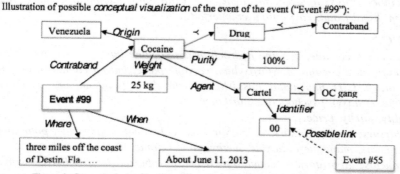

Figure 1: Conceptual visualization of the event. Other events may link to some of the attributes as illustrated for Event #55.

Fig. 13 Semantic analysis from a selected piece of information about cocaine trafficking (Henrik Legind Larsen, Aalborg University)

The next key step was the identification of key terms. A literature review facilitated several sources, like dictionaries and glossaries (i.e. Elsevier's Dictionary of Drug Trafficking Terms).

This is an example of a way to select key terms:

ACTORS
Producer countries: *Colombia, Peru, Bolivia...*
Origin countries: *Brazil, Venezuela, Caribbean Islands (Jamaica, Netherlands Antilles, Martinique)...*
Transit countries: *Guinea Bissau, Mali, Cape Verde, Madeira, Azores, Canary Islands, Benin, Gambia, Ghana, Guinea, Nigeria, Sierra Leone, Mauritania, Togo, Algeria, Libya, Morocco, Black Sea, Baltic Sea, Balkans, Spain, Portugal, the Netherlands...*
Destiny: *Portugal, Spain, France, the Netherlands, UK, Germany...*
Roles: *international wholesaler, national wholesaler, retailer, boss, manager, partner, transporter, tester, money collector, mixer, storer, law enforcement official, courier, money deliverer, drug abuser, drug addict, drug baron, drug dealer, drug peddler, drug runner, drug smuggler, hitman (sicario), networks, drug treatment,...*
Groups: *narcos, Mafia, 'Ndrangheta, cartels, FARC, AQIM, Galicia traffickers, Sinaloa Cartel, Zetas, Outlaw Motorcycle Gangs OMCGs, Hell Angels,...*

RELATED CRIMES
Drugs: *Cultivation, production, dispensing, distribution, logistics, marketing, frauds obtaining prescription drugs, processing, importation, manufacturing, possession (for own use, with intent to sell), sale, supply, distribution, trade, payment, recruitment, transportation, consumption, drug flows...*
Other crimes: *money laundering, other trafficking activities (weapons, human beings, other drugs, tobacco, medicines,...), extortion, kidnappings, smuggling, terrorism, fiscal fraud, corruption, counterfeit...*

PRODUCTS
Cocaine, crack cocaine, cocaine hydrochloride, cocaine base, coca leaves, cannabis resin, pure cocaine, Erythroxylum, cocaine paste, herbal cannabis, opium, heroin, crack, LSD, marijuana, hashish, synthetic drugs (amphetamine, methamphetamine, ecstasy), psychoactive substances,...
Quality, purity, price...
Precursors: *pre-precursors, precursor chemicals, acetic anhydride, potassium permanganate, manganese dioxide, methyl ethyl ketone, acetone, toluene...*
Chemicals: *laboratories, secondary extraction laboratories, HCl, HCl manufacturing, fertiliser, plastic, herbs, clothing, liquids, guano, beeswax, upholstery, polypropylene...*
Adulterants, cutting agents: *anaesthetics lignocaine (lidocaine) and benzocaine; painkillers such as phenacetin (a carcinogenic substance) and paracetamol; and other agents such as hydroxyzine, boric acid, glucose, manitol, lactose and caffeine, levamisole...*

TRANSPORT
Routes, Containers, couriers, mules, postal services, airports, digestive tract, ships, airplanes, aircraft, submarine,...

POLICE
Drug raid, squad, forensic science, asset confiscation, arrestees, entrapment, drugs seized, seizures, arrests, detentions, interceptions, wastewater analysis...

It is very important to point out the use of slang, as well as the need to introduce these kinds of specific terms to collect information and detect trends. Talking about cocaine:

Other cocaine names (popular names, slang, in different languages): *24/7, aspirin, Charlie, Carrie, Otoban, Dinamite, BLO, Big C, Diablo, Dios, Devil's Dandruff, Devil's Drug, Devil's Dick, Paradise, Polvere di stelle, Polvo Feliz, Polvo de Oro, Heaven Dust, Haven Dust, Happy Powder, Happy Trail, Dream, Beam, Soplo, Soddio, Angie, Gulosa, Ice, Icing, Flakes, Snow, Neive, Nieve, Snow White, Biancaneve, Bianca, Blanca, Blanche, Branca, Branquinha, Beyaz Ten, Caballo blanco, Belaia loshadi, White girl, white tornado, white lady, white dragon, white ghost white powder, polvo blanco, polvere bianca, poudre, pudra, sugar, azúcar, koks, cocco, coconut, coco, Coke, perico, perica, farlopa, calcetín, cama, Paco, Fefe, Bernie, Cecil, Baby, Bebé, Love Affair, Fast white lady, Lady C, Lady Caine, Dama blanca, girl, girlfriend, Mamá Coca, She, Her,...(from the book "CeroCeroCero", Roberto Saviano)*

4.2 Indicators

There are limitations with the identification of indicators:

- Criminal activity is not public and open, so it is not easy to collect data and information that could be evaluated: credibility and reliability.
- There are not good statistics and homogeneous data. Lack of data and poor quality.
- There is not academic consensus (or solution) about the way to measure organised crime, and usually indirect indicators are used (weapons without control, types of crimes, money laundering, black markets and economy, etc.)
- An only indicator cannot explain a phenomenon. A multivariable analysis is needed.
- Talking about cocaine trafficking, a key question is that there is an origin and a destiny, an offer and a demand, and the factors involved in both parts are not the same, and sometimes they are opposed.
- Temporal limitations. Indicators need to be built. Several times there is an important temporal gap between the moment in which we have the indicator and the moment it is referred to. At least, we have a static measure to compare (temporal dimension—year, month, week—and spatial dimension—countries, regions, cities).

Thinking about cocaine trafficking, this could be a preliminary approach:

Economic: *markets, prices, profits, drug demand, drug supply, economic crisis, inequality, poverty, economic freedom...*

Indicators: Economic Freedom Index (The Heritage Foundation), Unemployment rates, Gini index, GDP...

Political: *weak or failed states, corruption, geography (border control, mountains, ...), judicial system, armed conflicts, behaviour of elites, means shortage, political system, links with terrorism, violence...*

Indicators: Corruption Index (Transparency International), Democracy Index (EIU), Fragile States Index (FP) and its components, Global Peace Index, Global Terrorism Index, number of..., % of...

Social: *demography (age, sex, nationalities,...), migrations, education level, human development, unemployment, health system, urbanization, ...*

Indicators: urbanization trends, Human development index (UN), ...

Technological: *Internet, drones, deep web, Darknets, Black markets (Silk road) innovation, laboratories, online payments, bitcoin...*

Indicators: internet penetration, number of users of applications, laboratories detected, amount of payments on-line...

Legal: *legal system, judicial system, prison, ...*

Indicators: new legal frameworks alerts, number of prison population,

Crime: *age, sex, nationalities, type of crime, day of the week, hour, place, modus operandi, means used, media seized, ...*

Indicators: a set of criminal statistics.

5 Conclusions

Criminal phenomena are more a question of evolution and adaptation than novelty. Criminal groups easily identify the strategic and operational opportunities created by the environment and exploit "gaps" and "disconnects", taking advantage of the situation given its decentralised structures, the absence of norms and values in their actions, and the availability of large economic resources and specialised human resources (lawyers, experts in finance and taxes, money launderers, hackers, technology developers).

Modern societies face asymmetric threats, due to the flexibility of the criminal world. To tackle this threat, states and international organisations need to have consensus, compliances and fidelity to democratic principles, which limits their ability to anticipate and to fight against criminal actions.

Based on the previous analysis, we could assure that organised crime has an evolving environment that will facilitate its action and its development in the coming years. A higher degree of international disorder, new routes and economic opportunities, criminal globalisation, changes in markets and consumption habits, and technological advances will shape the future. This is a future in which the fusion between organised crime and other figures (terrorism, insurgencies) will grow.

For these circumstances, the analysis of the environment and trends is set as elements of strategic interest in the fight against organised crime. The broad lines on the present and future of the phenomenon are drawn. The aim is to obtain and analyse information in order to adopt an effective decision-making.

The fight against organised crime requires moving from risk management to opportunities management. The development of intelligence capabilities and the application of methodologies to improve knowledge of the criminal phenomenon, with the support of new technologies for information management, are the main tools for the detection and mitigation of the present and future criminal activity.

References

Blanco, J. M. & Cohen, J. (2014). The future of counter-terrorism in Europe. The need to be lost in the correct direction. *European Journal of Future Research*, 2(1), Springer. Retrieved May 10, 2016 from http://link.springer.com/article/10.1007%2Fs40309-014-0050-9

Bensoussan, E. & Fleisher, C. S. (2013). *Analysis without paralysis. 12 Tools to make better strategic decisions*. Pearson Education Inc.

Clark, R. M. (2013). *Intelligence analysis. A target-centric approach*. London: CQ Press.

Chatham House. (2015). *Growing threat as organized crime funnels radioactive materials to terrorists*. Retrieved May 10, 2016 from https://www.chathamhouse.org/expert/comment/growing-threat-organized-crime-funnels-radioactive-materials-terrorists

Council of European Union. (2015). *Serious and organised crime threat assessment 2017. Revised methodology*. Retrieved May 10, 2016 from http://www.statewatch.org/news/2015/dec/eu-council-socta-2017-methodology-14913-15.pdf

Development, Concepts and Doctrine Centre. (2014). *Global strategic trends out to 2045*. Ministry of Defence UK. Retrieved May 10, 2016 from https://www.gov.uk/government/publications/global-strategic-trends-out-to-2045

Europol. (2010). *EU policy cycle EMPACT*. Retrieved May 10, 2016 from https://www.europol.europa.eu/content/eu-policy-cycle-empact

Europol. (2013). *EU serious and organised crime threat assessment (SOCTA 2013)*. Retrieved May 10, 2016 from https://www.europol.europa.eu/sites/default/files/publications/socta2013.pdf

Europol. (2015). *Exploring tomorrow's organised crime*. Retrieved May 10, 2016 from https://www.europol.europa.eu/publications-documents/exploring-tomorrow%E2%80%99s-organised-crime

Grint, K. (2010). Wicked problems and clumsy solutions: The role of leadership. In S. Brooks & K. Grint (Eds.), *The new public leadership challenge*. Basingstoke: Palgrave Macmillan.

Goodman, M. (2015). *Future crimes. A journey to the dark side of technology—and how to survive it*. London: Penguin Random House.

Heuer, R. J & Pherson, R. H. (2014). *Structured analytic techniques for intelligence analysis*. CQ Press.

Lia, B. (2005). *Globalisation and the future of terrorism. Patterns and predictions. Contemporary security studies*. Routledge.

Makareno, T. (2004). The Crime-terror continuum: Tracing the interplay between transnational organised crime and terrorism. *Global Crime, 6*(1), 129–145.

Naím, M. (2013). *The end of power*. Basic Books.

National Intelligence Council. (2016). *Global trends 2030. Alternative worlds*. Retrieved May 10, 2016 from http://globaltrends2030.files.wordpress.com/2012/11/global-trends-2030-november 2012.pdf

Petersen, M. (2011). *What I learned in 40 years of doing intelligence analysis for US foreign policymakers. CSI studies* (Vol. 55, n° 1). Retrieved May 10, 2016 from https://www.europol.europa.eu/content/eu-policy-cycle-empact

Rand Corporation (2001). *Networks and Netwars. The future of terror, crime and militancy*. Retrieved May 10, 2016 from http://www.rand.org/pubs/monograph_reports/MR1382.html

RAND Europe. (2013). *Europe's societal challenges. An analysis of global societal trends to 2030 and their impact on the EU*. Retrieved May 10, 2016 from http://www.rand.org/pubs/research_reports/RR479.html

Shelley, L. I. (2014). *Dirty entanglements. Corruption, crime and terrorism*. New York.: Cambridge University Press.

Svendsen, A. D. M. (2015). Advancing "defence-in-depth": Intelligence and systems dynamics. *Defense & Security Analysis, 31*(1), 58–73. Retrieved May 10, 2016 from http://www.tandfonline.com/doi/full/10.1080/14751798.2014.995337

Svendsen, A. D. M. (2015). Contemporary intelligence innovation in practice: Enhancing "macro" to "micro" systems thinking via "system of systems" dynamics'. *Defence Studies, 15*(2), 105–123. Retrieved May 10, 2016 from http://www.tandfonline.com/doi/full/10.1080/14702436.2015.1033850

Taleb, N. N. (2013). *Antifragile: Things that gain from disorder*. Random House.

U.S Army Research Institute for Behavioral and Social Sciences. (2004). *Think like a commander*. Retrieved May 10, 2016 from www.dtic.mil/cgi-bin/GetTRDoc?AD=ADA425737

World Economic Forum. (2015). *The rising threat of organised crime on social media*. Retrieved May 10, 2016 from https://www.weforum.org/agenda/2015/07/social-media-violence/

World Economic Forum (2016). *The global risks report 2016*. Retrieved May 10, 2016 from http://www3.weforum.org/docs/Media/TheGlobalRisksReport2016.pdf

Extracting Future Crime Indicators from Social Media

Thomas Delavallade, Pierre Bertrand and Vincent Thouvenot

1 Introduction

Policing and fighting against criminality and terrorism can take different forms. First strategy relies on the combination of random patrol, rapid response and reactive investigation. Another strategy consists in building a strong link with the population, and work on partnership and prevention. Intelligence-led policing (ILP), a third strategy based on situation assessment and risk management, was first experimented in New York and in the UK. In New York, it was an initiative from commander William Bratton (NYPD) who managed to reduce crime in 1994. In the UK, it was used to identify chronic offenders, see for instance (Ratcliffe 2012). Yet, ILP became a popular strategy with 9/11, which changed significantly police strategy in the US and in the rest of the world and sparked a more widespread and smarter application of statistics to a more global issue: fight against organized crime. All of the different strategies introduced above produce data. Gathering all those data enables the development of a predictive policing model relying on new technologies, new business processes, and new algorithms with the intent to improve police efficiency.

As explained in (Beck and McCue 2009), the recent economic climate imposes budget reduction for many police departments which have to do more and more with less. Predictive policing allows to anticipate crimes and consequently a better management of police staff. Other activities, like marketing or competitive intelligence, developed efficient tools to predict and classify consumers' behaviors. Similar data analysis methods can be used for predictive policing. More details about this topic are given in (Pearsall 2010).

Data involved in predictive policing studies come from heterogeneous sources and allow many different analytical techniques, like data mining, crime mapping,

T. Delavallade (✉) · P. Bertrand · V. Thouvenot
Thales Communication and Security, Gennevilliers, France
e-mail: Thomas.delavallade@thalesgroup.com

geospatial forecasting or social network analysis. To be efficient, predictive policing needs to increase and diversify the number of data sources, and use some local data, such as demographic or sociological data. Data involved in predictive policing can be public, private, individual, confidential, and so on. They may be sensitive, which requires an ethically sound framework. This issue is studied in some European projects which are described later.

Fight against crime is a global issue. To deal with this, the European Union formed a European Law Enforcement Agency (LEA), Europol, which coordinates LEA's of the different member states in information exchange, intelligence analysis, and training. The 2007 Europol report (Symeonidou-Kastanidou 2007) provides a description of an always-evolving organized crime challenge. The report defines a list of dynamic crimes notably related to synthetic drugs, counterfeit goods, cybercrime and illegal immigration that are kindled by a globalized and organized criminality. Among other factors, open circulation of goods and persons, but also the Internet, have greatly contributed to the transnational development of organized crime groups. For instance, the Internet offers easy, instantaneous and almost free communication means but also commercial services on almost any kinds of good, be they legal or not, physical or virtual. Yet, some old-fashioned principles remain. For instance, as described in Drozdova and Samoilov (2010): *the least participants involved know about each other and their clandestine mission, the more survivable their social network is*. Hence there is a vital coordination-security trade-off: joint tasks demand coordination and communication, while these circumstances threaten organization's operational security and survival. To improve this trade-off, criminal organizations attempt to limit traces by safeguarding and scrambling communications. *"If this strategy potentially makes their direct and timely identification highly challenging in the operational environment, such attempts, however, necessitate communications technology choices that possess certain signature characteristics. These features can themselves be detected by contrasting with baseline usage patterns and so help reveal the increase in activities leading up to an attack"*. In Europol (2016) they provide statistics about the countries of residence, birth and activity of a suspect and notice that they can all three be different. These statistics show the evolution of crime toward a globalized organized crime. Hence, there is a need for police officers to share amounts of centralized information so that criminals could not snitch with missing data from a country to another.

Previous remarks raise two important issues. First, international data sharing methods but also processes between LEA's (Law Enforcement Agencies) will be key in the fight against organized crime and terrorist activity. Second, LEAs need tools to mine these data and extract from them features enabling the LEAs to better predict and anticipate crimes.

In this global context, Europe funds some international collaborative projects with a mix of LEAs, industrial and academic partners in order to favor European research on topics linked with the fight against organized crime and terrorism. SECILE (Securing Europe through Counter-Terrorism: Impact, Legitimacy and Effectiveness) for instance is one such project which addresses the impact, legitimacy and effectiveness of European measures to counter-terrorism. Other projects like IRISS (Increasing Resilience in Surveillance Societies) are more focused on the societal effects of criminality surveillance technologies. Privacy preserving technologies are also carefully considered in such projects. See INDECT (Intelligent information system supporting observation, searching, and detection for security of citizens in urban environment) for instance. Another important topic is the collaboration between European LEAs. One important side of collaboration is data sharing. CAPER (Collaborative information, Acquisition, Processing, Exploitation and Reporting for the prevention of organized crime, see Casanovas et al. 2014) aims for instance at developing a common information sharing platform dedicated to the detection and the prevention of Internet-related organized crime activity. As criminal organizations diversify their activities and increase their collaboration at European and global level, it is important to improve strategic early warning. The ePOOLICE[1] (early Pursuit against Organized crime using environmental scanning, the Law and IntelligenCE systems) project works on this topic and aims at building an efficient environmental scanning to enhance early warning.

All these European projects show two important topics in the fight against global crime. The information can be sensitive, so there is a need for a strong legal environment. Data need to be shared between police of different countries; hence, Europeans countries need to develop shared technologies, and to collaborate in the analysis and in predictive policing.

Traditional predictive policing approaches generally extrapolate spatiotemporal patterns observed in crime records, which are sometimes enriched with external open data about demography, economic situation, and so forth. For example, PredPol,[2] probably the most popular tool in this domain, allows crime prediction and mapping by using past crime type, place and time with an algorithm based on criminal behavior patterns. Eventually this tool provides the visualization of hotspots on a map. A lot of information is recorded during investigation. Authors of Oatley et al. (2006) deal with data mining of forensic evidence, predictive system and crime visualization with clustering methods. Police reports are another important type of data source. Authors of Chau et al. (2002) dealt with data mining in this context. These reports correspond to unstructured paperwork filled out by police officers. They are not easy to use for an automated machine analysis because of their lack of homogeneity and the grammatical and typology mistakes that require a higher and human level of interpretation. It could be useful to extract from these reports some information, for instance named entities such as person names or addresses, to help investigation and judicial analysis. To this end the authors used a neural network. The method achieved good results to extract names and narcotic drugs on a set of Phoenix police reports.

In some countries like the Netherlands, these police narrative reports and criminal records are locally stored and can be shared in a common nation-wide criminal database, such as a police national computer system. It is a rich data

---

[1]https://www.epoolice.eu/.
[2]http://www.predpol.com/.

source. Authors of de Bruin et al. (2006) apply data mining on a similar US database to classify past criminals. The study of these data allows authors to elaborate a new vision of what they name criminal career compared to more social studies based on traditional data. The four key factors to extract from such data are: (i) crime nature, (ii) frequency, (iii) duration, and (iv) severity. Authors use a multi-step process. The four key factors are first extracted, then criminal profiles of each individual are built by year, and finally differences between profiles are computed and visualized. They use clustering methods to identify usual criminal careers according to factors such as type of crime or number of crimes during the past year. Such a data-driven method provides automatic clusters enabling police officers to enrich expert profiles. It allows them to experiment with the validity of expert profiles. Moreover, they can easily compute new profiles according to incoming sources of data. Indeed, with new data sources and a traditional method, one would have to express new profiles according to new variables. With this method, profiles will be computed automatically so as to estimate the usefulness of new variables. Classifying criminal careers is not the only way to use local and nation-wide criminal databases. As said at the beginning, the 9/11 attacks led to major changes in the USA. Efforts have been done to collect digitally all crime and police reports at state and local levels. COPLINK (Chen et al. 2003), which is an integrated information and knowledge management environment, has been developed in this context and includes two modules. The first one, COPLINK Connect, allows for sharing in an easy way crime and police relevant content, while the second, COPLINK Detect, allows the use of police databases to detect and analyze different criminal associations.

However, there is a major issue encountered by traditional predictive approaches, which is the bias toward geographic areas well covered by existing police forces, because more data are collected and recorded for these areas. There is a need to build a more contextual and event aware model, which would rely on the previous standard model and extend it with indicators extracted from evenly distributed global data. Data have to come from multiple sources from private police files to open sources. Among open sources data, one can use social media data (Swendsen 2013). Links between individuals may enable police officers to extract intragroup communication patterns and to focus on the neighborhoods of these groups. Besides highly dynamic social media or micro-blogging platforms like Twitter can be used to alert when an unusual event appears thanks to emerging topic detection techniques (see Sect. 2.2.2 for more details on these techniques). Moreover, as people using such platforms can share a GPS position, one can focus on a given area.

In the following paragraph, we will focus on social network analysis, from the extraction to analysis and visualization which is a highly relevant research field to study social media feeds. In this part, keeping an eye on its relevance toward predictive policing, we explain how social networks can be analyzed in a more general context. social network analysis is an important topic with a large state of the art worthy of further articulation. Here, we briefly present three types of analysis. For further insights, interested readers can refer to Scott (2000). First, we consider the collection of news sources available via the Web, then we present two articles on Twitter studies, and finally we consider an example of a graph-oriented social network analysis.

Authors of Radinsky and Horvitz (2012) deal with the use of Web news to forecast future events, such as a cholera pandemic in Africa. The method employed in the article can be applied for the purposes of forecasting other issues. The underlying idea is to consider that real-world events and the associated news reports on the Web are generated by a probabilistic model. Authors extract digital news from the archives of 22 years worth of articles from the New York Times. Besides, they augment this corpus with Web articles from Wikipedia, FreeBase, OpenCyc, and GeoNames (see Linked Data project Bizer et al. 2009). They extract some sets of topically coherent news by using topic detection and tracking as presented in Cieri et al. (2000) and cluster similar texts together. Then, they extract Linked Data and define lexical and factual features for events and sequences of events. Finally, predictive models are computed.

Twitter, which is one of the biggest social networks, is another Web data source. Tweets are text messages published on Twitter and have a maximum length of 140 characters. More than one hundred million tweets are posted each day. It is a highly reactive platform which contains instantaneous user generated contents from all over the world that can be used for police prediction. Lots of research studies have been conducted to analyze this important knowledge source, for example for sentiment analysis (e.g., Go et al. 2009; Agarwal et al. 2011) or event detection (e.g., Chew and Eysenbach 2010 for pandemic study or Loten et al. 2011 for political events). In Teulf and Kraxberger (2011) they work on extracting semantic knowledge from Twitter. They propose a method based on Semantic Pattern concept (Teulf et al. 2009) and apply it to study the 2011 revolution in Egypt by making semantic clustering and relations. In Weerkamp and de Rijke (2012) they work on Dutch Tweets to forecast popular hobbies. A multi-step process is proposed, with: (i) selection of tweets, (ii) selection of key terms, and (iii) summarization of tweets.

The construction of indicators from open and closed sources has been performed by multiple researchers so as to demonstrate the usefulness of open source information to complement police files. For instance, in Williams et al. (2016) they examine the use of crime indicators based on open-source data. Conducting feature extraction from open data is actually a digital evolution of the "Broken Windows" indicator theory. A neighborhood degeneration leads to an increased probability of crime; it links visible forms of disorder to crime probability of occurrence. Polls or investigations historically allowed police officers to detect such visible indicators. Nowadays, tweets about a "Broken Window" may be available directly on the Internet. One just has to collect and analyze them. The paper shows a positive correlation between the frequency of such tweets and burglaries.

As mentioned in previous paragraphs about the way criminal groups are organized, communication is a key part of coordination especially for a global operation and criminals tend to use secured networks. Yet, as anyone, they are using social media in their private lives. Hence, monitoring social media communications may provide LEAs ways to reconstruct organized crime communities based on open interactions. For instance, using Twitter as an open database to predict a probability of crime has been performed in the two following papers: Bendler et al. (2014) and Gerber (2014). Both papers collect data from Twitter in order to estimate a crime probability in a given area and are also based on the "Broken Window" theory. The second paper constructs an indicator for each area in Chicago and each type of crime among prostitution, motor vehicle theft, kidnapping, etc. From historical crime records, they extract an average level of criminality per area and type. Then, using Twitter, they compute a current spread (as in finance terminology) above the average activity. To compute the spread, they estimate features for each type of crime (like the number of topics about burglary) that lead to increase the spread. Using GPS data from tweets, they can filter on a given area. Then, using these indicators they can detect risk areas and suggest the top priority areas where police forces should be deployed to prevent probable crimes. Another source of data was experimented in Iqbal et al. (2012) where they collected messages from online chat rooms and created cliques (fully connected communities) identified by common topics using clustering techniques. Then, one can filter on a type of topic and zoom on cliques that are linked to it. Hence, the main issue is to identify pertinent topics and cliques related to a given subject. Then alarms can be raised when a new topic-related clique appears.

As shown in the last papers, one can use Internet-related data like social media to detect, anticipate and even prevent real crimes like burglaries or thefts. Social media allow police officers to gather data from different international sources and create indicators about the geographic distribution of crime levels. These indicators depend on the type of crime and may be gathered from different open sources. The previous references show that police investigation and predictive analysis are evolving from a static knowledge of risk areas to a dynamic one based on both public and police data. Combining data from different sources leads to a more accurate crime modeling, allowing to focus on a specific area or type of crime. Open data provide an automatically updated source that can be combined with static knowledge to alert when there is an unusual event. Yet, long-term investigation can also be managed as exposed in Cohen (2013).

To reinforce the interest and importance of social media monitoring for crime prediction and prevention let us come back to the previously mentioned Europol study (Europol 2016), in which they present statistics about migrant smuggling traffic and notice that some advertisement may appear on social media, such as the post: "The cost of a package with travel from Turkey to Libya by air and onward sea journey from Libya to Italy costs USD 3700. For the sea journey adults cost USD 1000. Three children cost USD 500". These platforms are also used by migrant smugglers and irregular migrants to share information on developments along migration routes, including law enforcement activities, changes in asylum procedures, or unfavorable conditions in countries of destination. This type of information allows other migrant smugglers to adapt to changing conditions. Migrant smugglers adapt their pricing models in response to developments such as increased border controls by charging higher prices for alternative and safer routes. Hence, social networks may supply highly relevant information about criminal links and criminal activities.

The aforementioned studies all deal with crime indicators or the extraction of crime patterns in order to enhance crime prediction and above all prevention. While Europol studies provide trends and patterns at a strategic level, all the works mentioned in the domain of predictive policing are much more driven by operational needs. In both cases it appeared that taking into account social media feeds could enrich the analysis. However, most studies are confined either to the strategic or operational level. Besides, among those studies which included social media mining in their crime analysis, we observe two different approaches. Some focused only on the analysis of published textual contents, through an extraction of events and topics. Others focused only on the analysis of social media interactions, i.e., link (or relational) analysis.

In this paper we propose a common methodology to take into account social media for crime prevention at both strategic and operational levels, tooled-up by a common social media monitoring platform called OsintLab. Developed as a generic platform within Thales Big Data lab, it has been specifically customized for the fight against organized crime within the framework of the ePOOLICE project. It has besides been further enriched for anti-terrorism studies with French end-users. We first present this platform and its underlying methodology, before presenting the results of the experience gained thanks to using this platform. Discussing the corresponding experimental results, we will then be able to provide more generic statements about the types of indicators that can be extracted from social media. Finally, we provide future research directions in this field.

2 Methodology for Extracting Crime Indicators from Social Media

In this section we first give a brief overview of the OsintLab social media monitoring platform that has been used to perform the experiments described in the next section. We then introduce the key algorithmic and visual analytics components integrated in the platform. These components are the main strategic and operational enablers of OsintLab, upon which the intelligence-driven process we propose for crime indicator extraction is built. The formalization of this process is therefore presented as a synthesis of the technical description of OsintLab. Finally, before moving to the experimental results, legal and ethical issues related to the real-life deployment of such a platform within a LEA structure are outlined.

2.1 OsintLab Platform Overview

OsintLab is a social media monitoring platform which provides an end-to-end processing chain to collect, process, analyze, exploit and visualize in near real-time textual information published in social media like blogs, forums, Twitter or Facebook. Semi-automated, relying on highly interactive man-machine interfaces to allow analysts investigate large amounts of textual information, it aims at helping them to uncover potential threats and to understand information flow patterns. It is therefore an *investigation-oriented platform* rather than a *traditional monitoring platform* that would instead simply compute and display various kinds of statistics. Of course such statistics are also available and can be used to automate simple alerting tasks, but the real value-added of OsintLab lies in its ability to offer easy ways of exploring large and complex data streams while taking into account both textual and relational pieces of information. Thanks to this combined analysis and its visualization, users are able to easily understand what is being discussed, when, by whom and how influential these people are, including by which communities of users and how these communities are structured.

OsintLab relies on a modular, flexible and service-oriented back-end architecture as illustrated in Fig. 1.

Front-end components of this architecture combine a search engine, dashboards and a graph visualization in order to offer investigation support and various ways of extracting crime relevant indicators. Back-end components are combined in a generic workflow, based on a message passing pattern. This generic workflow includes by default standard social media components like crawlers, indexing and persistence components. Besides it integrates components dedicated to the analysis

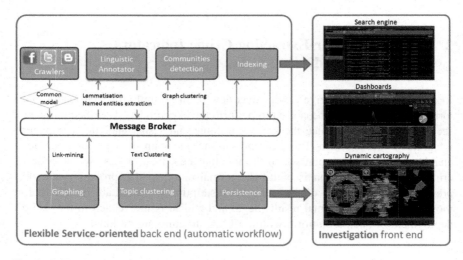

Fig. 1 OsintLab architecture

of text contents (linguistic annotator and topic clustering) plus components dedicated to link analysis (graphing and communities detection).

Thanks to its architecture OsintLab is a rather generic platform which can be used in various application domains in civilian, defense or homeland security markets.

For each target application domain, the generic architecture described in Fig. 1 may be adapted and specific components may be plugged in or out to match end-users requirements. For instance a GIS and specific cartographic search capabilities are added in a crisis management operational context. See Delavallade et al. (2016) for more details on this topic.

2.2 OsintLab Key Analytical Features

If OsintLab investigation capabilities are primarily realized through its visual analytics components, which are described in the next section, the relevance of these front-end capabilities heavily relies on some key back-end algorithmic components: a pattern-based text classifier, a text clustering engine and a graph clustering engine. We now present them briefly in the rest of this section.

2.2.1 Rule-Based Text Classifier and Alerting Mechanisms

Being able to automatically classify documents, based on their contents, within predefined semantic categories is a common yet crucial functionality of Web monitoring platforms. Indeed, it helps analysts to structure their knowledge base according to some categories of interest. Integrated within OsintLab, this component allows an analyst to define such categories and to link them with a specific pattern. Identifying texts matching these patterns can be done with neural-network-based natural language processing algorithms (Lai et al. 2015) or with a more linguistic approach (Gonçalves and Quaresma 2008). A pattern actually corresponds to an expert rule whose output is a predefined semantic category and which may take as inputs various criteria:

- text content: any keyword-based boolean query may be used
- metadata such as author of the text, source, publication date, crawling date, geolocation, etc.

Once such rules are defined, the collected stream of texts can be processed on the fly to assign each text to the categories for which it matches the associated patterns, following standard filtering.

In addition to the structuring of the knowledge base, this real-time, expert-driven classification mechanism is used as the basis of OsintLab's alerting mechanism. Indeed, the analyst can define meta-rules for each predefined category to be alerted as soon a peak of activity is observed in this category, i.e., as soon as the number of newly classified texts exceeds a user-defined threshold within a time window whose length is also specified by the analyst.

2.2.2 Text Clustering Engine

Cyber security signature-based intrusion or antivirus detection methods are mandatory not to miss known threats. Yet, they are not sufficient to fully cover more complex and evolving threats (David and Netanyahu 2015). Similarly, the rule-based classifier presented above offers an efficient way to track relevant issues, but only those which are known a priori to be relevant. We also need more data-driven approaches to allow for the discovery of surprising but potentially important issues, just like signature-based cyber security techniques are complemented by behavioral analyses based on anomaly detection.

Following the insights from this perspective, we have integrated a text clustering engine within OsintLab. It relies on the optimization of the Zahn Condorcet criterion (Conde-Céspedes et al. 2015) and tries to maximize the intra-class similarity while also maximizing the inter-class dissimilarity. This feature is the main advantage of our framework compared to popular inertia-based techniques like k-means which focus on the minimization of intra-class variability only (see Rajput et al. 2012 for more details about k-means used in the context of social media analysis). The trade-off realized by our approach leads to more stable results. Moreover, while k-means need some heuristics to choose automatically in a data-driven way the optimal number of clusters, and thus could be hard to apply when there are a millions of texts to cluster, this number is automatically chosen during the optimization process by our approach.

Furthermore OsintLab integrates a cluster summarization component which automatically labels the obtained clusters based on the five most discriminating terms of each cluster. A measure similar to the popular Tf-Idf (Spärck Jones 1972) has been chosen to assess the discriminative power of each term.

Within the ePOOLICE project in particular, OsintLab has been customized to capture among social media feeds, hidden signals of organized crime activities. The aforementioned topic clustering component which was primarily suited for strategic analysts has been extended to enable incremental analysis, i.e., analysts now have the possibility to perform "online" topic detection based on streams of data, rather than being forced to perform only batch analysis on historical data. This is useful to detect in near real-time trending and emerging topics.

2.2.3 Graph Clustering Engine

Social media feeds do not carry only information about topics and events through their textual contents, they also provide intrinsically relational information through user interactions. To exploit such information, we apply social network analysis techniques to automatically detect user communities, i.e., those groups of social media users which interact much more together than with users outside their community. This is achieved thanks to a graph clustering engine which relies on the Louvain algorithm (De Meo et al. 2011), probably the most efficient that can be found in the literature in terms of computation speed.

These types of communities are important to understand information flows within a given network and to highlight those users who drive information diffusion: community leaders, but also influence relays: those who enable information propagation between communities.

These considerations are highly relevant with regard to the operational needs of end-users. Indeed, they can be used to track some communities and understand their structure, which is of utmost interest when they are linked with organized crime activities, for example, or when we need to understand the information diffusion strategies of jihad propagandists.

2.3 Deep Investigation Using Visual Analytics

To exploit the information generated by the aforementioned algorithmic components, three visualization components have been designed and integrated into the OsintLab platform: two well-known and widely used visualization tools, namely a search engine and dashboards, plus a more unusual yet advanced visual analytics component based on a large graph visualization tool.

We will not give much detail here about the search engine and the dashboards, as they are very common decision support tools that can be found in any business intelligence application. To focus only on the most innovative features, we can simply mention that:

- the search engine integrates facets, in particular about the Web domains cited by the collected texts (for Twitter we perform first URL resolution in order to circumvent the tiny URL constraint);
- the dashboards provide various statistics about the collected **texts**, but also the **actors** who wrote them, the **sources** form which they have been published and the **topics** they address, as detected by the text clustering engine.

The third visual component has been designed specifically for deep investigation. It consists of a multi-graph view of the same data which are displayed from a statistical point of view in the aforementioned dashboards. As can be seen in Fig. 2, three graphs are displayed to show interactions between (i) texts, (ii) actors and (iii) topics; a timeline (iv) is also present at the bottom. All these four views are actually interconnected. Indeed, for each text we know (a) when it was published, (b) who wrote it and (c) which topic it addresses. These interconnections make it possible for analysts to easily search in one of the above four views (i–iv) and subsequently observe the impact on the three other views. The analyst has therefore

Fig. 2 Multi-dimensional investigation tool. *From left to right* we have the graphs of texts, actors and topics. A timeline is available at the *bottom*

at hand four entry points to dig into the data. Please note also that the automatically detected communities are displayed as colored convex hulls for both the actor and topic graph.

As can be ascertained from the appearance of the text graph on the left, this visualization tool is able to display very large graphs (more than 200,000 nodes and 500,000 edges in the above example) while ensuring fluid interactions. This is very rare and only possible because we developed this application on top of the TULIP,[3] framework, one of the very few which is able to reach such performances.[4] The ability to interact smoothly with the various graphs is a crucial point since simply visualizing all data would be useless given the volume of data to analyze. Therefore, this graph visualization tool has been designed to offer various filtering capabilities on the graphs themselves, based on node labels, node types, on the community to which they belong and also on the timeline.

Using the same visualization tool, we are able to display and link the topics and communities that have been extracted. We can for instance visualize which communities are active on a given topic and conversely which topics are being addressed by a given community. Hence, it is mainly through this tool that OsintLab manages to efficiently combine content and link analysis.

---

[3] http://tulip.labri.fr/TulipDrupal/.
[4] As a comparison measure, keep in mind that popular graphical frameworks like d3.js can hardly display a graph with 1000 nodes....

2.4 OsintLab Process for Intelligence Elaboration Support

Extracting crime indicators from social media can be seen as a specific intelligence task and more precisely as a specific OSINT (Open Source INTelligence) task. As such, the methodology followed to extract these kinds of indicators should rely on an intelligence process. Actually such a process is a cycle (see for instance Capet and Delavallade 2014 for more details) composed of the following steps:

(1) Planning and direction: formalization of the request for intelligence.
(2) Collection: it aims at orienting the sensors, in our case Web crawlers, in order to gather raw data which may help answer the request.
(3) Processing and exploitation: transformation of raw data into relevant information.
(4) Analysis: it mainly aims at evaluating the extracted pieces of information (assess information credibility and source reliability) and fusing them to elaborate pieces of intelligence.
(5) Dissemination: diffusion of these pieces of intelligence, in particular toward those who made the request for intelligence. The original request may then be refined and a new cycle begins.

Instrumenting all these steps is a highly challenging task and we have chosen, when implementing, the second and third steps, focusing therefore on raw data collection and extraction of relevant information. If data collection is a rather standard task, there are plenty of different solutions which have been proposed to build relevant pieces of information based on raw textual data. Within OsintLab, there are two analytical processes which are implemented to offer two different ways of uncovering hidden patterns:

- monitoring of issues known to be relevant with the combined classification/alerting mechanism described in Sect. 2.2.1;
- discovery of relevant but unexpected issues through visual investigation processes. This is made possible by aggregating the outputs of data-driven clustering components (see Sects. 2.2.2 and 2.2.3) within a visual analytics component dedicated to in-depth investigations (see Sect. 2.3).

Both analytical processes are essential for intelligence analysts. They are complementary as the first one enables to raise alerts according to a predefined expert threat model with a high precision (low false alarm rate) but potentially low recall (we may miss a lot of important issues). On the contrary, knowledge discovery allows to detect potential threats with a high recall (we catch much more actual threats) but with a low precision (we may generate lots of false alarms). By combining both processes, it becomes easier to find the right balance between false alarms and silence (all those true alarms that were missed). A common process consists in regularly updating and extending the expert classification rules according to the knowledge discovered thanks to the data-driven approach.

In Fig. 3 we sketch the methodological process implemented thanks to OsintLab.

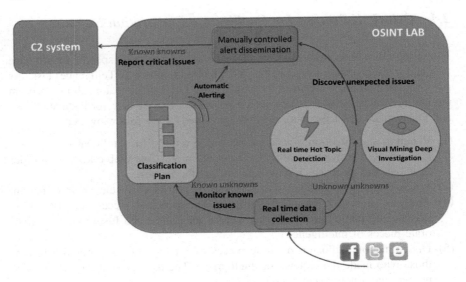

Fig. 3 Intelligence-driven analytical process instrumented by OsintLab

In this figure, the issues which have been identified as relevant are transferred at a higher level for decision support. This actually can be assimilated to the previously mentioned dissemination phase of the intelligence cycle. Note furthermore that we use a wording familiar to intelligence analysts in the context of knowledge management which has been popularized by Donald Rumsfled in 2002, while he was the U.S. Secretary of Defense:

> As we know, there are known knowns; there are things we know we know. We also know there are known unknowns; that is to say we know there are some things we do not know. But there are also unknown unknowns—the ones we don't know we don't know

This terminology primarily expresses the fact that intelligence has to face two main kinds of uncertainties while elaborating pieces of intelligence. Firstly, we must identify if and when some issues that are known to be relevant, with respect to the intelligence request, do occur. It corresponds to what we called a monitoring and alerting task. From a statistical point of view, this is similar to hypothesis testing. Secondly, we must have ways to uncover relevant issues even if we are not initially aware that they could be relevant to answer the request for intelligence. This corresponds to a knowledge discovery task.

2.5 Non-technical Considerations

We have described in the preceding sections how OsintLab, from a technical point of view, could enable analysts to implement a rather generic intelligence process to extract knowledge from social media, and undertake those tasks in particular about criminal and terrorist activities. However to implement such a methodology in practice, other dimensions should also be taken into account.

Firstly LEA's organizational and operational procedures should be adapted to take into account the new opportunities offered by OsintLab-like platforms. Without such organizational adaptations, it is likely that some important functionalities will be neglected and even worse that these tools remain unused. In a crisis management context for instance, we have tried to assess the usefulness of social media for PPDRs (Public Protection and Disaster Relief organizations) in the iSAR+ project.[5] Live experiments revealed that, by default, information extracted from social media are left aside by operational teams during intervention until they are confirmed by emergency calls. This results from the fact that PPDR current rules of engagement completely ignore such media which are therefore considered as unreliable by default.

Another important issue concerns personal data protection. Indeed, any document collected from social media contains such personal data. From a legal and ethical point of view, it is mandatory to consider a privacy preserving policy while deploying social media monitoring platforms for LEAs. Hence, during OsintLab development and evaluation phases, anonymous or fake data has been used. Within the ePOOLICE project, for instance, we have defined an ethically sound framework. At the strategic level, analysts are operating beyond merely individual-related personal data and it would be possible to automatically transfer them some suitable and aggregated indicators extracted from social media. However, for more operational and judicial purposes, some LEA's analysts need to perform deep investigation and access the original, un-anonymized data. In order to satisfy both requirements while minimizing the exposure to personal data leakage, we have chosen to isolate OsintLab from other analytical components developed by ePOOLICE partners. To do so we have adopted a loose integration framework (consider the link between OsintLab and an external C2 which is described in Fig. 3) so that only predefined messages related to indicators that end-users have decided to monitor can be transferred out of OsintLab. With this solution, we avoid the "contamination" of the whole intelligence chain, since only OsintLab analysts are in the end facing privacy preserving issues. This is a risk minimization solution but of course specific measures should be set up for the OSINT platform itself. They concern not only the security (physical and cyber) of the infrastructure hosting this platform to avoid any sensitive data leakage but also organizational processes and human resources management. Indeed, specific training and clearance for OSINT analysts should be given.

---

[5]http://isar.i112.eu.

3 Experimental Evaluation

To illustrate the usefulness of the methodology described previously, we provide in this section some concrete examples of relevant indicators that we managed to extract from social media in two distinct use cases. The first one deals with copper theft, an activity in which organized crime is known to be involved and which has a strong impact on our economies as will be explained in more depth in Sect. 3.1. The second one deals with jihadist propaganda and is presented in Sect. 3.2.

For both use cases we present some examples of crime indicators that we managed to extract from Twitter feeds using OsintLab and that may be relevant for consideration by both strategic and operational level analysts.

3.1 Copper Theft Use Case

Copper theft consists in stealing items for the value of its constituent copper. Hence, it follows the movement of copper price. Yet, during the last ten years, the copper price has fluctuated from around 1200 USD per ton in 2000 up to 10,000 in 2011. Now, copper price is around 4000. Hereafter, we will describe a short analysis of some indicators we can extract using OsintLab. We introduce two types of indicators according to the targeted main end-users: strategic or operational analysts. In the next two subsections, we present the two kinds of indicators we were able to extract with OsintLab by presenting the results obtained on a small study of Twitter activity related to copper theft reports. The Twitter search API has been used, with copper theft related keywords, in order to do a one-time collection of a total amount of just over more than 6000 tweets.

3.1.1 Strategic Indicators

Extracted indicators can reveal some trends and global statistics that may be used by authorities in order to watch over and manage the traffic. For example, copper theft does have a strong impact on two main industries: train traffic and electrical networks. Indeed railways and electric wires provide a not-always watched-over metal source. Therefore, electrical installations and railways represent good opportunities for copper thieves. Theft in these two kinds of infrastructures may have severe consequences on the quality of the associated services with a potentially high number of impacted citizens, such as various businesses' commuting employees and customers. It is, however, possible to closely monitor the effects on these two types of services on social media.

As an illustration we provide in Fig. 4 the trending topics about copper theft detected by OsintLab in the right window and so the associated tweets in the left window. In this figure we have focused on the impact on electric networks by

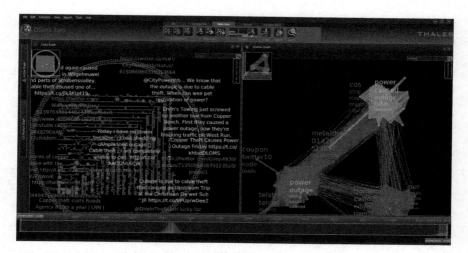

Fig. 4 Power outage detection using OsintLab

highlighting power outages related topics. Note that the impact of copper theft on train traffic could also have been shown. Indeed, various topics among the trending ones deal with train delays sparked by copper theft.

Combining this information with geolocation[6] information would be useful from a strategic point of view as it would allow for a direct and real-time assessment of the geographic distribution of copper theft according to the types of attacked infrastructures.

We can also compose some global indicators about international context that may cover for example economics and legislation. Indeed, using OsintLab, one can follow up copper price fluctuations. For instance, we detected a current decline of copper price in China that may impact the international market and therefore change the cost/benefit assessment that any copper thief may more or less consciously make: "Why are China's copper scrap imports in long-term decline?". In the same tweet, a link to a newspaper report was also given. Information about copper price could be found by a more standard method; yet, our approach is able to raise alerts. Besides, it comes with three advantages: (i) it links information from different sources and subjects, (ii) it detects interesting evolution without any a priori knowledge and (iii) it prevents us from manually following various subjects.

Legislation evolution is another topic emerging from OsintLab copper theft study as illustrated by a few tenths of tweets mentioning regulation change in the UK and in South Africa. This kind of information is difficult to integrate quickly in an LEA dataset, and it is therefore highly valuable to have ways of extracting almost automatically some meaningful PESTEL (Political, Economic, Social,

---

[6]This information could be extracted from Twitter metadata when present or extracted from text content with named entity extraction tools.

Technological, Environmental and Legal) indicators that would be otherwise cumbersome to gather together in the same information system.

Apart from global indicators, OsintLab allow detecting local trends which can be used both for a strategic and an operational overview. For instance this tweet about a reward offered by a company: "Verizon offering rewards for copper thefts information." It demonstrates the economic damages from which companies like Verizon suffer and it defines an area where copper thieves have to be less risk averse to remain active.

3.1.2 Operational Indicators

Twitter enables communications for both legal and illegal copper markets. Hence, it is common for scrapyard owners to advertise that they are buying copper metal. While such an announce is not suspicious in itself, it may also be a public façade for illegal transactions. Detecting those communications provides a useful indicator which may help to track local areas where opportunities to resell stolen copper are higher. This can be useful to improve situation awareness. For instance, in Fig. 5 we can see tweets from buyers and sellers in London. Such tweets supply operational information by defining actors and geographic areas to watch over.

Moreover, OsintLab, through the detection of emerging topics, can provide operational insights to uncover previously unknown and therefore unmonitored threats. Emerging topics are the topics automatically identified by OsintLab which are not yet very popular but are being discussed more and more. For example, information about damages in scrapyards may be linked with an illegal activity like copper theft. Hence, it is aggregated in an emerging topic by OsintLab which refers

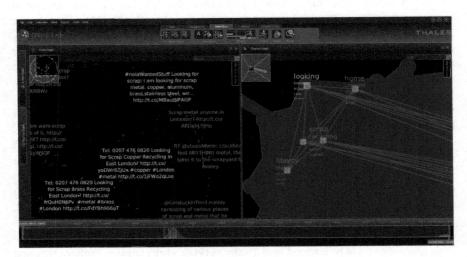

Fig. 5 When copper demand and offer meet on Twitter

to the event described by the following tweet: "Pueblo scrap metal yard will face charges after two fires". Detecting these kinds of events can be seen as a decision support information when it comes to allocate police forces. Finally, they can also follow up the arrest of copper thieves in a given area to check how strong the associated criminality is. An important number of tweets like the following one similarly report these kinds of events as well: "Two men arrested for alleged theft or copper wire worth $50,000."

3.2 Jihadist Propaganda

Terrorist groups like Al Qaeda and even more ISIS (Islamic State in Iraq and Syria) have fully integrated in their global strategy information war concepts. They have their own news agency—Aamaq for ISIS, they publish magazines, like *Dabiq* or *Inspire*, in various languages and are able to finely segment their target audience and to adjust the messages (both content and form) for each target group. They have highly professional communication agencies which are able, in particular, to produce high quality videos. They master information diffusion processes on social media such as Twitter, Facebook, Tumblr, etc., and are even able to develop high-tech tools to ease the coordination of their information campaigns like the Twitter app "The Dawn of Glad Tidings". Among other goals, these campaigns aim at:

- reassuring potential allies w.r.t. their control of the local situation, their confidence in the victory;
- frightening the home populations of states involved in the war against them in order to influence a democratic change against the governments fighting them;
- discrediting rival groups;
- recruiting new fighters.

The numerous terrorist bombings and attacks in various places around the world and in particular the two consecutive attacks in Paris and Brussels at the end of 2015 and the beginning of 2016 have pushed the fight against terrorism and consequently the fight against jihadist propaganda at the top of homeland security departments priorities. Many research studies have been and are being conducted to assess the prevalence of such a phenomenon in social media (Berger and Morgan 2015). A Kaggle[7] has even been posted on this topic.

Before presenting our own experimental results, we need to describe briefly the data collection process we have set up for this use case. We used the Twitter streaming API and subscribed to a few keywords and accounts that were manually selected beforehand. Then we used OsintLab knowledge discovery tools in order to regularly update the list of keywords and accounts to follow. This iterative update

---

[7]https://www.kaggle.com/crowdflower/first-gop-debate-twitter-sentiment.

of the monitored keywords and accounts is crucial since Jihadist accounts are frequently censored by Twitter and new ones are constantly being created. Through OsintLab graph visualization analysts are being informed of which accounts have been closed and which new accounts are being frequently cited. Besides, we used these first analyses to identify some relevant issues for which we set up dedicated classification rules and alerts. The collection process has unfortunately not been in place continuously but only for a few days at three different periods: just after the 2015 November attacks in Paris, in December 2015 and in February and March 2016 after the Brussels attacks. Eventually we managed to compile a corpus containing around 178,000 tweets with the following language distribution: 30,000 in French, 34,000 in English and 114,000 in Arabic.

3.2.1 Strategic Indicators

This static view of the language distribution is an interesting indicator in itself, but from a strategic point of view it would probably be more interesting to monitor the evolution of this distribution in order to detect relevant trends. Activity bursts in a given language can indeed reveal a successful recruiting effort by propagandists or the development of an efficient new cyber tool in a specific linguistic community involving the deployment of large botnets to relay their activity. On the other hand, a sudden activity drop could result from successful cyber to physical operations, such as raids and arrests, conducted by LEAs in a given country.

At the same strategic level, statistics about the most cited Web domains is important to understand the trends with respect to the use of content hosting platforms. It allows to quickly figure out the type (text, audio, image, video) of illegal content which is the most widespread. In addition, if this information was combined with geolocation information—where these platforms are physically hosted—it could help to target LEAs' collaboration efforts with the countries hosting the platforms which are the most popular among jihadists. To illustrate this point we provide in Fig. 6 screenshots of OsintLab giving the most cited Web domains per language.

In this figure, it is interesting to note that *justpaste.it* is by far the most popular content hosting platform. However, behaviors are slightly different when we compare the ratio of tweets mentioning this platform in all three languages: 93% in Arabic, 17% in English, 3% in French. In Arabic the exceptionally high number may be explained by the fact that "justpaste" which very quickly turned out to be often used by jihadist supporters was added as a keyword in the crawling while no specific Arabic keyword was included. So it is a matter of collection bias. Note, however, that we can verify that it is truly popular since, even without Arabic keywords, we managed to collect much more tweets in Arabic than in other languages. The comparison is therefore more relevant between French and English since both French and English keywords were used to seed our Twitter crawlers.

If we leave *justpaste.it* aside, other popular domains are Video hosting platforms like *Youtube, Dailymotion, Sendvid, vid.me* but also generic cloud storage platforms

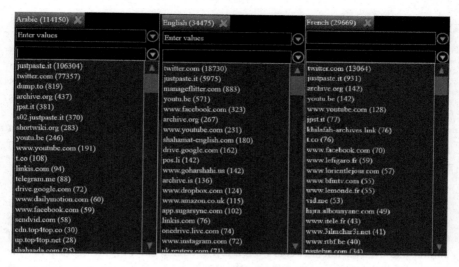

Fig. 6 Most cited web domains per language (Jihadist propaganda use case)

like *google.drive* or *Dropbox*. For Arabic, *Telegram*, the encrypted chat platform well known for its anonymity preserving properties, is well represented.[8] Although not visible in Fig. 6 for French and English, it is also quite frequently cited in these two languages. The observed citation ratios in our corpus are pretty similar, though a bit higher in English: 0.09% in Arabic, 0.15% in English and 0.09% in French.

At a macro level it is also relevant to get an idea of the geographic trends. For this we can use the geolocation metadata provided by Twitter, when Twitter users have accepted to share their geolocation. The corresponding indicators can also be scanned language by language to figure out if there are any peculiar habits depending on the linguistic community, as suggested by the authors in Leetaru et al. (2013). For example, in our dataset, we have only 49 tweets which are geolocated out of 178,294, which corresponds to a 0.027% ratio. This is extremely low. In Leetaru et al. (2013) authors mention an average ratio between 1 and 2%. This may be explained by the fact that Twitter users we are monitoring are more careful about their anonymity. Interestingly, English-speaking users are more prone to use geolocation since we observe that among the 49 geolocated tweets 43 were written in English, 3 in Arabic and 3 in French. From an operational point of view, this means that there are probably more opportunities to target precisely English-speaking users.

After these first findings extracted from the global statistics, we have used the rule-based classification process described in Sect. 2.2.1 in order to monitor the various issues known to be relevant: news about Brussels attacks, diffusion of video

---

[8]In the first months of 2016, Telegram eventually decided to censor lots of accounts from jihadist supporters.

contents and discussions around security, anonymity on the Web. Figure 7 gives the detailed statistics for all these issues.

Analyzing in detail all this data and what it can communicate is not within the scope of this chapter. Therefore, we now focus on a single sub-corpus of analysis to present the remaining relevant indicators but the approach is generic. This sub-corpus is referred to as 'French1' in Fig. 7: it contains all those collected tweets which are written in French.

At the strategic level, analysts may be interested by statistics about the most popular discussion topics. Thanks to OsintLab's text clustering engine, we can provide some popularity measures like:

– Number of tweets dealing with the topic
– Number of different actors discussing the topic

The latter is important because some "big" topics in terms of tweet numbers may be artificially big due to a strong bot activity: a bot may indeed publish lots of similar and even identical tweets from a few accounts only. Figure 8 illustrates the interest of these measures in our French corpus.

Among these topics, two are actually highly relevant for strategic analysts: the first one has been labeled "retour censure…" ("back censorship…" would be the English translation but the true wording we observed in our dataset, for English tweets, is "back suspension") and the second "compte secours…" ("account backup…" in English). They correspond to an important activity of jihad propagandists on social media which aims at overcoming censorship: it is actually a network reconstruction activity. Jihad supporters promote either their new accounts a posteriori, once they are back from suspension under a new account, or proactively their backup accounts when they consider censorship as imminent. Although such discussions may concern other kinds of censored accounts, given the keywords tracked on Twitter in our study, the likelihood that such accounts are somehow involved in terrorism apology is very high. Note that such a piece of information is also useful for operational analysts as it provides highly relevant ways of finding true jihadist supporters accounts on Twitter.

In a similar perspective, it is highly relevant to quantify globally the impact of censorship by looking at the closed accounts. As accounts may be closed either voluntarily by the owner itself, for example in order to change identity on the network, or censored by Twitter, it is interesting to distinguish both measures. This is now possible since Twitter has introduced in 2016 a specific HTTP error code for censored accounts. Monitoring the evolution of both the number of closed and censored accounts, normalized by the total number of accounts, can provide valuable strategic insights. For example, if we consider only the tweets written in

Corpus	Creation mode	Source types	Creation date	#Texts	#Actors	#Sources
tor	SEARCH_ENGIN	TWITTER	03/23/2016 15:25	294	247	18
english1	SEARCH_ENGIN	TWITTER	03/24/2016 08:30	28543	12951	212
french1	SEARCH_ENGIN	TWITTER	03/24/2016 23:35	25924	4212	68
all tweets	SEARCH_ENGIN	TWITTER	03/23/2016 15:42	166790	36334	257
belgium1	SEARCH_ENGIN	TWITTER	03/25/2016 18:14	1456	606	37
belgium1 weak signals	GRAPH	TWITTER	03/25/2016 19:02	649	279	21
citation of video hosting domains	SEARCH_ENGIN	TWITTER	03/23/2016 14:20	2118	1241	31

Fig. 7 OsintLab statistics about monitored issues

descriptors	#Posts	#Comments	#Actors
maroc hijra	11	100	96
islam	60	41	82
terrorisme apologie	23	94	81
kouffar	36	48	67
kippa voile	24	69	65
musulmans	44	54	64
bruxelles attentats	39	67	60
sifaoui	8	77	60
retour censure	29	32	55
magazine	102	1	52
messager allah	25	27	52
sbs allah	43	40	50
merci	47	16	49
compte	36	33	48

Fig. 8 OsintLab statistics about the main discussion topics

Arabic on one side and in French on the other we observe in our corpus the following ratios:

	Censored accounts ratio (%)	Closed accounts ratio (%)
French	9	14
Arabic	50	18

From these ratios we can observe that the Arabic accounts are much more censored than the French ones. This can be explained by the way we configured Twitter crawlers, i.e., we were much more accurate in terms of our targeting in relation to the choice of accounts and keywords to follow for Arabic compared to French tweets. Indeed, a large part of our collected data turned out to be actually linked to jihadist propaganda since half of the Arabic accounts in our database have been censored in the end. An alternative explanation would be that jihad supporters are hiding less in Arabic than in French or are more pushy in their support demonstration. To our mind, the first explanation should be privileged at this stage of the analysis. Indeed Arabic content has been targeted in our experiment mainly through the use of "justpaste.it" which is known to be extensively used by jihad supporters. French contents on the contrary were collected based on additional keywords, which may have been more ambiguous as they may be used by jihad supporters, hunters or analysts. This was useful to identify more diverse jihadist supporters, but it probably brought some noise in the analysis.

Other interesting indicators, more technical, concern the source application from which tweets have been published. For example it may be *Twitter for iPhone* or *Twitter for Android* or *IFTTT* (one of the many applications which can be used to automate the publication of tweets). This metadata is indeed publicly available and can be used to check what are the most popular applications and devices used by jihadist supporters.

Relational indicators may also be strategically valuable in order to quantify the size and strength of the information environment in which jihadist supporters are playing. To this end, thanks to OsintLab graphing and community detection components, it is possible to monitor the following indicators:

- Number of communities of a predefined size: this indicator gives a first idea of how concentrated the information environment is; for example in the French corpus there are 6180 actors spread in 70 communities, while if we consider the tweets written in Arabic there are 163 communities for 20,969 actors. So at first sight it seems that the Arabic social media users are a bit more concentrated if we consider the average community size: 88 for French users against 129 for Arabic users.
- Overall entropy of the information space computed as follows: $H(C) = \sum_{i=1}^{n} P(c_i) \log_2(P(c_i))$ where n denotes the number of social media users and c_i denotes the community to which belongs user i. This indicator refines the previous one in the sense that it quantifies the "impurity" of the classification of users in the different communities. It is maximal and equals $\log_2(n)$ if all social media users are evenly distributed in the different clusters. It is on the contrary minimal and equals 0 if all users are gathered only in one class. Surprisingly with this indicator we obtain very similar results with a 4.12 entropy for French users and 4.16 for Arabic users. So, contrary to what the previous indicator suggested, the diversity in terms of communities in both cases, French and Arabic, is similar. Monitoring this indicator could help for instance to detect an entropy decrease which could be interpreted as a strengthening of the jihadist networks. The closeness of both entropy figures is quite surprising. It would be worthwhile to figure out if this is purely coincidental, if this is a common feature of Twitter jihad propagandist network that can be observed in other languages, or if this an intrinsic property of the Louvain algorithm.
- Other relevant indicators to assess the strength of the jihadist networks are the interaction ratio (number of interactions divided by the number of actors), average and standard deviation of the different communities "connectedness". Here, we simply take the sum of the interactions in which the actors of a given community are involved, normalized by the total number of actors. The following table recaps these numbers for French and Arabic users, respectively. It turns out that jihadist supporters in the French Twittosphere are interacting more than in the Arabic one, with more compact communities.

	Interaction ratio	Average connectedness	Standard deviation connectedness
French	2.44	0.07	0.22
Arabic	2.27	0.028	0.132

3.2.2 Operational Indicators

While at the strategic level analysts need trend indicators, at the operational level indicators allowing to spot emerging topics are more relevant as we explained in the *copper theft* use case. Besides, specific indicators are required to target specific accounts involved in jihadist propaganda, potentially to initiate a judicial procedure. In this perspective we present in this section more actors-centric indicators.

Some indicators available in OsintLab dashboards can provide first elements to quantify the importance of the different actors in order to quickly identify those which deserve a closer attention. We have for instance built indicators to reflect the following actor characteristics:

- their **activity**: number of published tweets;
- their **influence**: number of citations received, number of citations received over number of published tweets ratio;
- their **social activity**: number of actors with which they interacted.

Figure 9 lists the most important French-speaking actors according to the number of tweets published. Note that due to the sensitivity of our dataset, actor pseudos have been anonymized.

Combining the activity indicator with an influence indicator like the number of citations received over the number of tweets published ratio, we can observe that user 309720 is not only very active but he is also very influential, much more than user 317617. However, had we presented the contents published by this actor, we

Pseudo	#Citations	#Ratio	#Posts	#Comments	#Actors
309720	3163	327%	967	1	477
317617	211	11%	591	1267	385
310866	252	56%	407	41	105
310602	1819	535%	340	0	633
310051	351	110%	279	41	148
312283	332	105%	263	53	152
315220	455	81%	254	309	264
311087	285	131%	186	31	78
228059	117	74%	151	7	82
323275	183	124%	148	0	24
310297	211	89%	142	94	173
309014	186	116%	137	23	75

Fig. 9 Actor-centric indicators in OsintLab dashboards

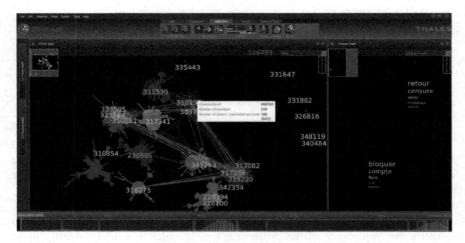

Fig. 10 Most active communities on topics related to jihadist network reconstruction (before or after accounts are suspended)

would have hardly observed any obvious relation with jihadist propaganda. They indeed mainly concern Islam art and history. Therefore, these indicators are important but not sufficient for an efficient targeting.

To enhance decision support at the operational level, OsintLab can be used to build indicators based on the combination of content and link analysis. The multi-graph visualization tool indeed provides a mapping between the detected discussion topics and the detected actor communities which is well suited for deep operational investigations. Based on this mapping we can extract statistics about the most active communities on a given topic. For instance, we identified in the previous section that some topics were related to jihadist network reconstruction. If we consider the most active communities for these topics and their leaders, then we have valuable information in an operational targeting perspective. Figure 10 illustrate this mapping.

On this figure we visually observe that only 3 or 4 communities are really active on this topic. They correspond to those communities having a significant number of highlighted actors. Among these communities, the one for which we display statistics is actually led by user 309720. This is the same user we mentioned previously. He is one of the most influential but looking only at the published contents was not enough to label him as a jihadist supporter. Now that we add *relational information*, evidence about the implication of this user in French jihadist ecosystem is much stronger. Besides evidence is even stronger if we take into account the ratio of *censored accounts* within a community. We can indeed observe in Fig. 10 that more than 1/3 of the actors belonging to the community led by user 309720 have been censored.

Figure 11 provides an additional example of the same capability. The highlighted topics were selected based on the keyword "photo". They are all interrelated

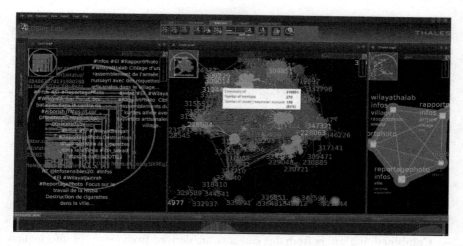

Fig. 11 Communities relaying ISIS reports and pictures from the field

as they all deal with reports (containing pictures) from the field of the life within ISIS caliphate. Only two communities are really active on these topics and we can observe that for the most active one, led by user 310051, half of the members have been censored. We can imagine that such a community should be carefully monitored as it disseminates in the French Twittosphere most of the news and pictures from the field. Some of the community members are therefore in touch with people from the field and may be involved in recruitment.

Extracting indicators combining content and link analysis and providing the associated visualizations is the true strength of OsintLab from an operational and investigation perspective. It offers an easy way to target the most suspicious communities and their leaders on a given topic.

Conversely, we can also derive relevant statistics about the most discussed topics per community and monitor such indicators to detect when a given community extends or narrows the scope of addressed issues.

4 Discussion

In the previous examples, we conducted an analysis over a couple of datasets. Both were generated by filtering Twitter streams based on manually selected keywords and accounts. Hence, while interpreting the various statistical indicators presented in the previous section, it is highly important to take into account the corresponding selection bias. In the description of both use cases we tried to clarify, whenever it was required, this selection bias. To discriminate between what results from this bias and what truly reveals a relevant risk factor deeper investigations are required. It is important at least to keep in mind the existence of this bias.

In the description of our two use cases, in the previous section we have introduced numerous indicators with a first classification attempt in two categories depending on whether they were relevant for strategic or operational analysts. Hereafter, we intend to generalize our findings. From this perspective, we introduce a second level of classification to differentiate indicators based on the type of data they have been derived from. More precisely, we consider for this second classification level the following categories:

- **Content**: indicators from this category are based on publication contents
- **Metadata**: indicators from this category are based on metadata associated with the collected texts (GPS localizations, authors and their publication habits, publication sources...)
- **Link**: indicators from this category are based on links between individuals. In our case we considered only interactions as links but follower/followee relationships could also have been considered
- **Combined**: indicators from this category are derived from pieces of information extracted from at least two of the three previously mentioned categories (content/link or content/metadata or metadata/link)

Besides this taxonomic effort, we intend to generalize the indicators extracted from our two specific use cases. This work is done in order to have the ability to easily derive relevant indicators for new use cases. The following list describes the main families of indicators we managed to identify in this generalization effort:

- **Volume**: high-level indicators about the global publication activity in a given use case, like the total number of documents, of citations....
- **Emerging topics**: indicators about topics which are still quite "small" in terms of size but whose popularity is growing fast.
- **Topic distribution**: volume indicators for the identified discussion topics.
- **Language distribution**: volume indicators for the different languages used in the studied use case.
- **Geographic distribution**: volume indicators per geographic areas.
- **PESTEL**: volume indicators related to Political, Economic, Social, Technological, Environmental or Legal factors, like for instance regulation changes.
- **Censorship**: indicators about censored actors.
- **Actor-centric activity**: indicators related to actors activity levels, like for instance the number of written texts, the publication frequency, etc.
- **Actor-centric influence**: indicators related to the influence of the different actors like for instance the number of citations received, the number of actors with whom a given actor has interacted, and so forth. Page rank-like indicators could also be considered as relevant members of this family.
- **Source-centric activity**: volume indicators for the different publication sources.
- **Community-centric censorship**: indicators about the level of censorship within a community.

- **Information space concentration**: indicators used to quantify the level of concentration of the information space associated to the studied use case. The overall community diversity based on entropy which has been introduced earlier is a paradigmatic example for this family.
- **Per topic community activity**: indicators related to the activity level of the different communities on a given topic.
- **Per community topic activity**: indicators related to the diversity of the topics addressed by a given community.
- **Competing communities**: indicators about the inter-community level of competition within the studied information space. Such indicators could be for instance derived from clustering approaches applied on the different communities based on their "topic profile".

Based on the proposed taxonomy and indicator families, we propose to summarize the main findings of our work about the extraction of crime indicators from social media in the form of the following table. Assigning an indicator to an application level, strategic versus operational, was not easy since as we observed in our case studies some indicators are often valuable for both levels. To avoid having all indicators in both levels which would have been a bit confusing for the reader, we eventually decided to associate each indicator only with a single application level and we chose to keep track only of the level for which we, a bit arbitrarily, assumed that it would be the most useful.

	Strategic level	Operational level
Content	Volume Topic distribution Language distribution PESTEL	Emerging topics
Metadata	Censorship Geographic distribution	Actor-centric activity Source-centric activity
Link	Information space concentration	Actor-centric influence Community-centric censorship
Combined	Competing communities	Per topic community activity Per community topic activity

5 Future Work

Many of the indicators presented in the previous sections are static: they rely on the current state of collected data and do not take into account any evolution. It could be interesting to develop dynamic indicators, taking into account trends and seasonality of studied phenomena. Any static operator can become dynamic when it is applied to a continuous set of collected data. This is exactly our case on Twitter for which we have the ability to collect continuously streams of tweets. Therefore, for all the previously mentioned static indicators, we can build the corresponding time series and derive dynamic indicators by looking at the evolution and trends of these time series. For instance, we may extract time series features such as growth rates, or moving averages.

A second improvement deals with the detection of anomalies and ruptures in static (e.g., outliers detection) and dynamic indicators (e.g., detect a peak of tweets about a topic or a topic change in a community). For time-series-based indicators, the common shape of the curve (integrating seasonality trends) could be learnt in order to detect outliers more efficiently.

In the previous sections, we did not deal much with geolocation. Indeed, only a few tweets were geolocated. As explained in the jihad use case, this may be due to some bias linked with the studied topic: jihadist supporters are aware of the dangers of cyber traces, like geolocation metadata, left on the Internet. Yet, for other studies, it can be used to construct improved indicators enabling police officers to detect a precise risk area. In order to circumvent the lack of precisely geolocated data, we could get inspiration from some authors who are using a relational analysis to propagate the known localization information and increase the number of geolocated Twitter accounts (Jurgens et al. 2015).

In this section we presented some important research directions which would be worth investigating further in order to improve the indicators we managed to extract in our two use case studies on copper theft and jihadist propaganda. The list is far from being exhaustive, but it gives an overview of our future work.

References

Agarwal, A., Xie, B., Vovsha, I., Rambow, O., & Passonneau, R. (2011). Sentiment analysis of Twitter data. In *Proceeding of workshop on languages in social media* (pp. 30–38).
Beck, C., & McCue, C. (2009). Predictive policing: What can we learn from Wal-Mart and Amazon about fighting crime in a recession? *Police Chief Magazine, 76*(11).
Bendler, J., Brandt, T., Wagner, S., & Neumann, D. (2014). Investigating crime-to-twitter relationships in urban environments—facilitating a virtual neighborhood watch. In *Proceeding of ECIS 2014*.
Berger, J. M., & Morgan, J. (2015). *The ISIS Twitter census defining and describing the population of ISIS supporters on Twitter*. The Brookings project on U.S. relations with the islamic world.
Bizer, C., Heath, T., & Bernes-Lee, T. (2009). Linked data, the story so far. In *IJJWIS*. http://linkeddata.org/docs/ijswis-special-issue
Capet, P., & Delavallade, T. (2014). *Information Evaluation*. Wiley-ISTE.
Casanovas, P., Arraiza, J., Melero, F., Gonzalez-Conejero, J., Molcho, G., & Cuadros, M. (2014). Fighting organized crime through open source intelligence: Regulatory strategies of the CAPER project. In *Proceeding of JURIX*.
Chau, M., Xu, J., & Chen, H. (2002). Extracting meaningful entities from police narrative reports. In *Proceeding of 2002 Annual National Conference on Digital Government Research*.
Chen, H., Zeng, D., Atabakhsh, H., Wyzga, W., & Schroeder, J. (2003). COPLINK—managing law enforcement and knowledge. *Communications of the ACM, 46*(1).

Chew, C., & Eysenbach, G. (2010). Pandemics in the age of Twitter: content analysis of tweets during the 2009 H1N1 outbreak. *PLoS One, 5*(11).

Cieri, C., Graff, D., Liberman, M., Martey, N., & Strassel, S. (2000). Large, multilingual, broadcast news corpora for cooperative research in topic detection and tracking: The TDT-2 and TDT-3 corpus efforts. In *Proceedings of Language Resources and Evaluation Conference*.

Cohen, L. (2013). *6 ways law enforcement uses social media to fight crime*. http://connectedcops.net/wp-content/uploads/2010/04/6-Ways-Law-Enforcement-Uses-Social-Media-to-Fight-Crime.pdf

Conde-Céspedes, P., Marcotorchino, J. F., & Vienne, E. (2015). Comparison of linear modularization criteria using the relational formalism, an approach to easily identify resolution limit. In *Proceeding of EGC-AKDM*.

David, O., & Netanyahu, N. (2015). DeepSign: Deep learning for automatic malware signature generation and classification. In *IEEE Proceeding*.

de Bruin, J., Cocx, T., Kosters, W., Laros, J., & Kok, J. (2006). Data mining approaches to criminal career analysis. In *Proceedings of the Sixth International Conference on Data Mining*.

De Meo, P., Ferrara, E., Fiumara, G., & Provetti, A. (2011). Generalized Louvain method for community detection in large networks. In *Proceedings of the 11th International Conference on Intelligent Systems Design and Applications*.

Delavallade, T., Fossier, S., Laudy, C., & Lortal, G. (2016). On the challenges of using social media for crisis management. In *Fusion methodologies in crisis management*.

Drozdova, K., & Samoilov, M. (2010). Predictive analysis of concealed social network activities based on communication technology choices: Early-warning detection of attack signals from terrorist organizations. *Computational and Mathematical Organization Theory*, (16).

Europol. (2016). Migrant smuggling in the EU. In *Europol public information*.

Gerber, S. (2014). Predicting crime using Twitter and kernel density estimation. In *Proceeding of Decision Support Systems*.

Go, A., Huang, L., & Bhayani, R. (2009). *Twitter sentiment analysis*. http://www-nlp.stanford.edu/courses/cs224n/2009/fp/3.pdf

Gonçalves, T., & Quaresma, P. (2008). Using linguistic information to classify Portuguese text documents. In *Seventh Mexican International Conference on Artificial Intelligence, MICAI '08*.

Iqbal, F., Fung, B., & Debbabi, M. (2012). Mining criminal networks from chat log. In *Proceeding of International Conferences on Web Intelligence and Intelligent Agent Technology*.

Jurgens, D., Finethy, T., McCorriston, J., Tian, X. Y., & Ruths, D. (2015). Geolocation prediction in Twitter using social networks: A critical analysis and review of current practice. In *Proceedings of the 9th International AAAI Conference on Weblogs and Social Media (ICWSM)*.

Lai, S., Xu, L., Liu, K., & Zhao, J. (2015). Recurrent convolutional neural networks for text classification. In *Proceedings of the Twenty-ninth AAAI Conference on Artificial Intelligence*.

Leetaru, K., Wang, S., Cao, G., Padmanabhan, A., & Shook, E. (2013). Mapping the global Twitter heartbeat: The geography of Twitter. *First Monday, 18*(5).

Loten, G., Graeff, E., Ananny, M., Gaffney, D., Pearce, I., & Boyd, D. (2011). The revolutions were tweeted: Information flows during the 2011 Tunisian and Egyptian revolutions. *International Journal of Communication, 5*.

Oatley, G., Ewart, B., & Zeleznikow, J. (2006). Decision support systems for police: Lessons from the application of data mining techniques to soft forensic evidence. *Artificial Intelligence and Law, 14*(1).

Pearsall, B. (2010). Predictive policing: The future of law enforcement? *NIJ Journal, 266*.

Radinsky, K., & Horvitz, E. (2012). Mining the web to predict future events. In *Proceeding of WSDM'13*.

Rajput, D. S., Thakur, R. S., Thakur, G. S., & Neeraj S. (2012). Analysis of social networking sites using K-mean clustering algorithm. *International Journal of Computer & Communication Technology*.

Ratcliffe, J. H. (2012). *Intelligence-led policing*. Willan Publishing.

Scott, J. (2000), *Social network analysis*. SAGE Publications Ltd.

Spärck Jones, K. (1972). A statistical interpretation of term specificity and its application in retrieval. *Journal of Documentation, 28*(1), 1972.

Swendsen, A. (2013). Introducing RESINT: A missing and undervalued "INT" in all-source intelligence efforts. *International Journal of Intelligence and Counter Intelligence, 26,* 4.

Symeonidou-Kastanidou, E. (2007). Towards a new definition of organized crime in the European Union. *15 European Journal of Crime, Crime Law & Criminal Justice,* (83–103).

Teulf, P., & Kraxberger, S. (2011). Extracting semantic knowledge from Twitter. In *Electronic participation* (pp 48–59). Springer.

Teulf, P., Payer, U., Parycek, P., Macintosh, A., & Tambouris, E. (2009). Automated analysis of e-participation data by utilizing associative networks, spreading activation and unsupervised learning. In *Proceeding of 1st International Conference on Electronic Participation* (pp. 139–150).

Weerkamp, W., & de Rijke, M. (2012). Activity prediction: A Twitter-based exploration. In *Proceeding of SIGIR 2012 Workshop on Time-Aware Information Access.*

Williams, M., Burnap, P., & Sloan, L. (2016). crime sensing wit big data: The affordance and limitations of using open-source communications to estimate crime patterns. *British Journal of Criminology.*

Organised Crime, Wild Cards and Dystopias

José María Blanco and Jéssica Cohen

> The illiterate of the 21st century will not be those who cannot read and write, but those who cannot learn, unlearn, and relearn. People will not be trained based on permanent knowledge in their minds, but in terms of their ability to know what is needed every time
>
> Alvin Toffler, 1971.

1 Imagination and Creativity to Manage Transnational Security Risks

The current complexity of combating organised crime is just one example of the challenges that will have to be faced in future. The complexity of the environment, encouraged by VUCA conditions (volatility, uncertainty, complexity and ambiguity,) transforms chaos into a new order.

Although the current situation is already complex, the most confusing scenarios have not arrived yet. The variables that define the present times have been repeated throughout history, with modifications in the context in which they occur or how they are combined. But, applying the mentality of a turkey (meaning that the treatment that it receives every day prevents it from being aware that it will be the main dish in Thanksgiving dinner) to contexts of increasing complexity does not seem to be the most appropriate approach. An event that has never existed is not impossible today, and imagining the failure becomes the first step to warn about it and prevent it.

J.M. Blanco (✉)
Centre of Analysis and Foresight, Guardia Civil, Madrid, Spain
e-mail: jmblanco@guardiacivil.es

J. Cohen
International Security Intelligence Analyst, Private Sector, Madrid, Spain
e-mail: mail@jessicacohen.es

© Springer International Publishing AG 2017
H.L. Larsen et al. (eds.), *Using Open Data to Detect Organized Crime Threats*, DOI 10.1007/978-3-319-52703-1_9

In 1971, Alvin Toffler analysed these variables to describe the symptoms of what he called the *Future Shock,* a context that, by pure definition, would be extremely harmful and painful for rigid and intolerant people and organisations. The present requires us to talk about the past and the future simultaneously, terrorism and organised crime on the same strategy, and prevention and eradication with the same intensity. Present times demand new tools and skills to tackle threats and generate opportunities that are still unknown.

A requirement result of this type of context is the need to direct our thoughts to the future, turning this action into a cognitive habit. Design all possible dystopias and scenarios, analyse the stakeholders, interests, variables, constraints and actions that could compose them and, based on this detailed study, start thinking actions to avoid any of the negative scenarios or to reduce their impact. Discard excessive confidence in what we know and concreteness, giving equal importance to not known situations and the general contexts in which criminal phenomena may occur. Recover the healthy habit of discussing, blank sheet and teamwork. Distinguish between what is urgent and what is important. This is neither easy, nor impossible.

The growing challenge posed by crime, like many other current variables, suggests the need to develop increasingly flexible and creative responses. And, in parallel, this also enhances the way we relate to others, how to interact with machines and how to manage the exponential increase of information available.

The challenge is to extend this need not only within the institutions and people that shape them, but also within the whole society, and to internalise that the value of intellectual capital is the sum of both current and potential knowledge. As Goodman (2015) explains in his latest book, *"Future Crimes"*, even if we are in the best possible situation where the institutions have the capacity, flexibility and skills to tackle crime derived from new contexts, the fight will be lost if society is not trained in its prevention.

In this situation, we must also ask ourselves what we understand as knowledge, including the knowledge related to a specific problem such as organised crime, and what we believe will be the knowledge in the future. A report from Oxford published in late 2013 augured a period of about 25 years for half of the jobs across the USA to be automated. So, our future perhaps is not to compete with machines, trying to be carriers of large quantities of information, errant drives, robots with repetitive tasks or data processors. For those who work against organised crime groups, from either the exploratory stages or academic research to the most operational tasks aimed to detect and prevent the criminal phenomenon, this change is a new challenge to the profession and needs an analysis from within. As Gerd Leonhard defends, we may need to unlearn the habit of acting like machines and relearn to act as human beings.

Two of the latest publications of the World Economic Forum with the Boston Consulting Group (2015, 2016) are demanding these changes, giving a detailed list of needs to be covered by the new academic approaches, in order to prepare adults to act with solvency in the new environments, and to train them in the management of the risks that are involved but still unknown. This results in a response focused in three pillars, as follows:

(i) *Be the challenge: internalisation of change.*

Non-traditional threats require greater solidarity between the different actors who have to face them, the internalisation of lessons learned, the consideration of more and new fields of action, and an extraordinary effort for flexibility and adaptation in response. And these characteristics imply the need for a new-style education model that enables new skills for which we have not been trained so far. This education should emphasise that a personal redefinition must be constant and that the current knowledge and skills are becoming obsolete.

The future is not always a projection of the past and the present. Rupture events modify the possible linear evolution that our minds can draw. There are risks with low probability and high impact, the so-called wild cards. We do not know what will happen, the information is limited, "both concerning the behaviour of others who are oriented to us and also concerning the future that will result from our choices", so we have the opportunity to revise this "present future" in the continuous "future presents". Esposito, in this situation, thinks that time is a great opportunity and that it allows for imagination and creativity.

And in the case we had the best possible information, new tasks will be continuous duties: how to evaluate its veracity, classify and reclassify it, how to establish links between data, categorise and relate it back, how to focus on details and later in the context again and how to analyse the problem and the direction in which it manifests itself. In summary, how to learn, unlearn and relearn as an adaptive response.

(ii) *Creativity as a habit: the control of complexity.*

We need a creative and imaginative approach in order to anticipate new risks. Imagining, listening, experimenting, making mistakes, creatively creating and destroying, using intuition, applying emotional, cultural, and social intelligence, or making unexpected connections are key skills for living in the future and tackling the challenges that lie ahead. Knowledge is becoming a set of skills, not an accumulation of knowledge. The goal will be to create a differential value through a particular skill at a given time.

The report about 9/11 stressed in a chapter devoted to prospective analysis ("Foresight and Hindsight") that the lack of imagination was the major mistake when trying to prevent terrorist attacks. Imagination is not common in bureaucracies. Policy-makers have a short horizon, but policies should not. *Unknown unknowns*, a famous concept popularised by Donald Rumsfeld, are perhaps impossible to manage. Taleb (2012) creates the concept of "antifragility", a way to manage the uncertainty (wild cards or black swans) without trying to identify them previously, but through preparation to be stronger in this evolutionary and complex system when events with low probability happen.

According to the 9/11 Commission Report "it is therefore crucial to find a way of routinising, even bureaucratising, the exercise of imagination". Richard Clark (National Counter-terrorism Coordinator, NSC, 1997–2001) attributed his awareness about the possible use of airplanes as weapons more to Tom Clancy novels than to warnings from the intelligence community.

Imagination is needed to identify new crimes and attackers, to discover vulnerabilities, to think about new modus operandi, to connect the dots (one of the key functions of intelligence analysis), to preview scenarios, to establish hypothesis to evaluate, to suggest different alternatives, to have different points of view and to develop new ways and new processes of analysis.

(iii) *Networks of people: the effectiveness of the community is the effectiveness of its members.*

A report published in 2008, "Knowledge Tools of the Future", stated another requirement in this evolution, which is the need for highly creative individuals, groups and networks: Family and community environments that encourage experimentation, risk and error; institutions (both public and private) who can accommodate to these new attitudes and ways of doing, leaving behind the anachronism of the closed structures, overly hierarchical or authoritarian which limit the growth of human capital; governments acting in line with the complexity of the contexts and the priority of the problems. All actors must understand that not thinking in terms of future implies a lack of strategies.

2 The Need to Study the Future in Criminal Intelligence

There are several difficulties with uncertainty, complexity and change dynamics. The methodological effort would try to combine a positive knowledge about future trends with a lack of transparency. We deal with the distinction between a "present future" and a "future present". Esposito (2011), following the sociological systems theory of Niklas Luhmann, points out that the past and the future include a multiplicity of past and future presents. Even if the world is unknown, one still has a known and acceptable orientation about the future. We know emerging trends that could be consolidated in future, and we know future events (elections, demographic evolution) and technological developments that will be points of change.

Game changers could be managed through a monitoring system of information, analysing the evolution of each one. Wild cards are impossible to detect and incontrollable. It is possible to get a methodically vision of "present future", through trends analysis, horizon scanning or other techniques (tools to understand the future, but not scientific methodologies). But we do not know what will happen, the information is limited, "both concerning the behaviour of others who are oriented to us and also concerning the future that will result from our choices", so we have the opportunity to revise this "present future" in the continuous "future presents". Esposito, in this situation, thinks that time is a great opportunity and that it allows for imagination and creativity.

Foresight, in criminal intelligence, could be:

- A sensor that alerts about change processes that could occur in all areas (political, social, technological, economic, legal, environmental, communications and transport), and affect, either positively or negatively, the criminal phenomenon. Detecting trends and identifying indicators can facilitate the identification of weak signals and the configuration of early warning systems that warn against changes in criminal patterns or the emergence of new risks and threats.
- The only known way that could avoid acting as a reaction after the materialisation of new risks.
- A way to expand the range of actors in the security system, which increasingly includes the private sector, the media, universities, think tanks, associations, communities and minorities, or citizens themselves. There is no state monopoly on knowledge on security (Blanco 2015).
- An input to update the view of organisations that have among its missions the fight against criminality and organised crime. The organisational vision determines how they intend to achieve their mission. The vision must be adaptable to the environment. And that environment can only be analysed through holistic models and in collaboration with the rest of the actors in the system.
- The study of the future provides:
  - A different way of thinking, based on creativity and critical thinking.
  - A reflection on future risks and threats.
- Thought and reflection on the future introduce causality in the system, conditioning the becoming future.
- A needed input for strategic design.

3 Methodological Approach

Several techniques and tools can help to create this vision of the future from the future (and not from the present): wild cards, scenarios, backcasting, what if, dystopias or the collection "Futures Research Methodology" from The Millennium Project (2009) (United Nations University).

Trying to simplify theoretical questions, we can identify two lines in future studies applied to criminal intelligence, the first one would have the objective of prediction (forecasting) and the second an objective of prospective (foresight).

(a) Predictive intelligence: forecasting.

Clark (2013) identifies three possible approaches: extrapolation, projection and prediction.

Extrapolation predicts the future assuming that the variables that have influenced the object of research will not vary. It has a very-short-term vision. It is useful in stable environments in which the variables are well known. Correlation, a measure of the degree of association between two data sets or the extent to which two variables are related, would be the main tool. Regression is a technique to determine the value of a variable based on the current value of other variables. Multivariable regression involves the use of more than one independent variable to predict the value of a dependent variable. Through extrapolations, we arrive to a single future scenario.

The projections assume a change in the variables that explained a phenomenon in the past and present, but without introducing new variables. Therefore, they present several future scenarios and assign probabilities to each. It may be useful in short and medium terms, but not long-term, or in environments of rapid change.

In recent years, there has been widespread use of the concept "predictive police" (for more information, read Chapter "Foresight and the Future of Crime: Advancing Environmental Scanning Approaches"), which refers to a set of methodologies used by the law enforcement agencies, aimed at implementing strategies to prevent the commission of the offence or conduct investigations more effectively (Rand Corporation 2013), enabling:

- Predicting crimes. You can determine locations and times, when and where new crimes will be committed.
- Predicting the possibility that an individual commits a crime. Techniques are being developed to predict the human criminal behaviour, whose source could be even analysis of language used in a social network.
- Predicting groups and individuals who may be subject of the crimes.

To achieve these objectives, it is possible to use several methodologies and techniques: analysis of crime data, regression models, classification and clustering, geolocation of information, geographical profiling and analysis (conditions of places that encourage crime), the establishment of patterns, criminal profiling, early warnings (any deviation from a pattern), statistical modelling, data mining, big data (mass data management), advanced algorithms or artificial intelligence.

Based on the positive assessment of the combination of methodologies and technologies applicable to predictive testing, we must point out its limitations:

- It is not possible to know the future. Technology is not able to give a solution. Computers are not "crystal balls". Living in a VUCA world, with many actors taking decisions continuously, it is impossible to know what is going to happen in a very short time. Wild cards or black swans will happen, introducing disruptive elements in the future.
- It is not possible to automate all processes. Technology only supports them, being needed the action of human analysts. Technology should be a tool, not an objective.

- Models need resources, both human and technological. Emerging technologies have a short life cycle before being replicated by competitors, so its cost is usually high.
- It is not certain that its use reduces crime. It is a support tool, an information resource and, in the best case, a simple prediction.
- It is essential to have good data. Poor quality data, inhomogeneous, or incomplete, reduce the quality of the prediction.
- Predictive systems can tell us what happens, but not its causes. Strategic management needs to identify causes to tackle.
- Overconfidence in the data. Intelligence analysis has always been characterised by the need to face uncertainty with incomplete information. Intelligence analysis experts, such as Lowenthal (2013), have been warning that the obsession with the data and information overload (infoxication) can paralyse the analyst.
- Overvaluation of predictive systems. These systems are based on past data. Therefore, they will likely detect events that have already existed. In this way, they can hardly cope with the so-called black swans (black swans or wild cards), events with very low probability but high impact.
- Limitation creativity, as a result of excessive reliance on data. Creativity is a pillar to design future scenarios.
- Privacy. Data acquisition and processing precise compliance with a legal minimum. Big data, as recognised by the US government (2014) itself, is a threat to citizens.
- The use of new technologies produces non-desired effects, leaving traces of the access to information by researchers and investigators.

(b) Prospective intelligence: foresight.

Current environments are characterised by rapid change, enormous complexity and uncertainty. Foresight is the study of the future to influence from the present. Foresight models try to offer future scenarios in order to adopt present actions that will lead to the desired future, or at least avoid the most damaging.

There are methodologies and techniques that are not only predictive but more oriented to the development of future scenarios and with a purely strategic management vision.

These models need to abandon the linear conception of time, "bureaucratising" imagination and creativity, a way to introduce these skills in the daily processes of our organisations. The management of chaos (Clark 2013), especially in a VUCA world, needs to involve specialists in many disciplines, diversity of views and strategic thinkers to build up the "big picture" of new criminal phenomena.

Social sciences provide different models, which combine complex data and incorporate new data to evaluate the present or predict the future (Box–Jenkins method), used usually in engineering (Kalman Filter). Clark's proposed model, based on the past and present, determines the combinations of variables that can lead to future scenarios, combining variables with no change, variables that change, or new variables.

Foresight techniques can be configured as a cohesive framework of diverse and fractionated knowledge, providing a holistic view from different disciplines about the criminal phenomenon. We do not believe in its value as a report or a static prediction, but as an ongoing monitoring and evaluation system that allows corrections on the designed future scenarios (Blanco and Jaime 2014).

For these purposes, we propose a combination of the following methodologies, to be applied by teams of analysts:

3.1 What if

What if analysis imagines that an unexpected event has occurred and it has a potential major impact. After that, the analyst or the team of analysts develops a chain of argumentation to explain how this outcome could have come about, in a backcasting process. For each possible scenario, analyst should generate a list of indicators that could help to detect whether events are starting to play out in a way envisioned by that scenario. Finally, indicators must be monitored on a regular or periodic basis (Heuer and Pherson 2014).

3.2 Wild Cards

A wild card is an event or future development with a low probability of occurrence but high impact. Facing the future, there are three types of possible surprises:

- Known events, with some certainty about their occurrence but not about the moment they will happen (i.e. the next earthquake).
- Unknown events to the general public, or even researchers, but that could be discovered if appropriate experts will be consulted or correct models were used (i.e. impacts of climate change).
- Unknown events, without preconceptions or means for observation (unknown unknowns).

Petersen (1997) developed a model for the analysis, beginning with the investigation of possible disruptive events that could happen. First, for that purpose, we should apply creativity techniques, expert interviews, questionnaires or historical analogies. Science fiction is a source of information. Second, we must evaluate the impact of these events on the subject of study or our organisation, with criteria such as vulnerability, the period in which it materialises, the existence of opposition to the occurrence, their involvement to the essence of the person, impacts and reaction rate (which will be lower the more unpredictable is the event). Third, potential triggers should be monitored, identifying facilitators and inhibitors, as well as indicators. These precursors act as "weak signals" and could be integrated into a model of "early warning".

3.3 Scenarios

The classic model, developed by Godet and Durance (2011), provides a set of methodologies to reach the construction of future scenarios. To do this, based on a preliminary analysis on the target subject or organisation, Godet proposes the identification of the key variables (MICMAC methodology), the analysis of the possible game of actors (MACTOR methodology), the exploration of the field of possible futures (morphological analysis, Delphi surveys, Regnier Abacus or cross-impact method), and evaluation of strategic options (relevance trees and MULTIPOL methodology).

There are other scenario generation models. One of the most famous ones was proposed by Schwartz (1996) in the Royal Dutch/Shell. It identifies four phases for creating a scenario, similar to those of any troubleshooting process: problem definition, identification of factors affecting the problem, proposed alternative solutions and finding the best option.

Scenario methodology has been used to analyse the future of terrorism (Gordon), but not usually for organised crime. However, there are several references in intelligence studies (Svendsen 2012a, b).

4 Dystopias

We understand dystopia, an antagonistic term of utopia, as a representation of an unwanted future, an undesirable hypothetical society. We opted for this concept compared to more restrictive definitions, and gird over the possible content thereof, as it considers the dystopia as a futuristic and imagined universe in which an oppressive social control is maintained and there is an illusion of a perfect society through a totalitarian, bureaucratic, corporate, technological or moral control.

The creation of the term is considered to correspond to John Stuart Mill, who in 1868, in a speech, used the term dystopia as an antonym of utopia.

In short, dystopias are scenarios, overly negative, which contribute to a criticism of trends, both political and social or technological.

4.1 Dystopias and Foresight Analysis

Some current trends are represented in texts of both analytical and fictional character: continuous conflicts, lack of transparency, corruption, crisis of governance, loss of confidence in institutions, uncontrolled growth of cities, increasing fractions of territory without control from the states, restriction of public freedoms and fundamental rights in the interest of greater security, inequality (economic, social and technological), migration, resource scarcity, terrorism (complex definition), extremisms brandishing one or another flag, increasing technological developments in the service also illegal actions and individual empowerment.

These issues, although not in such a dystopian way, are reflected in prestigious international analysis of countries and organisations that think about future. In this vein, the National Intelligence Council of the United States is preparing its new Global Trends 2035 report, which is published on the occasion of the appointment of a new president. It will replace the report of 2030. Moreover, the European Strategy and Policy Analysis (ESPAS) recently published its "Global Trends to 2030: Can the EU meet the challenges ahead?" report.

The value of these documents is not only their predictive ability, but also his contribution as early warning, which enables the evolution of trends through indicators and design strategies to limit the threats and opportunities. A task, so far, developed by countries and institutions that know what they want and how they want and that act to build your desired from this future. This is a slope for other organisations and states lacking strategic thinking culture and future needs.

Dystopias differ from disruptive events in the sense that they bring together a number of circumstances portraying a society of the future with a broader view that the so-called wild cards, referred to very specific events. They are complete scenarios of the future.

Gaston Berger, prospective guru, said that "foreseeing a catastrophe is conditional in that it requires one to foresee what would happen if we do nothing to prevent it" (Berger 1967). "Observing an atom modifies the atom, watching a person affects the person and looking at the future transforms the future" (Berger 1964). In his own words, "looking at the future disturbs not only the future but also the present".

4.2 How to Use Dystopias for Intelligence Analysis and Foresight

For these purposes, two phases are identified: construction and deconstruction. The construction phase is designed by futurists intended to reflect a possible situation, encourage debate, determine causal lines and enhance the generation of indicators on drivers and trends. The goal is to design a model that allows, in its phase of deconstruction, the generation of additional ideas and in-depth analysis about scenarios.

(a) Construction phase
1. Avoid the projection of current trends. For that purpose, there are other methodologies, techniques and tools, such as trend analysis and trend evaluation of trends, scenario building or analysis of variables and actors.
2. The sources of information to build a dystopia are the same as those identified in the methodology of disruptive events: news, science fiction, video games, official reports, etc. Taking particular importance in this case those inputs of information derived from the application of creative techniques and imagination.

3. Identify, for the construction of dystopias, the fundamental fears of a society, both those who respond to more objective threats and those more subjective or irrational. One possible source would be surveys, as the one published by The Chapman University, which as a major contribution provides a cluster in 10 domains of fear: environment, daily life, technology, natural disasters, personal future disasters caused by humans, government, crime, personal anxieties and judgment of others.

    While it may be a simple task, it is considered that current fears could not be the most important in future. But events and factors from the present, under other conditions, could be threatening in the future.

4. Identify the major risks and threats to humanity, trying to build future scenarios through them, especially through their combination. Several analyses provide guidance, for example the annual Global Risks (World Economic Forum 2016), Global Trends 2030 (US intelligence), or the Worldwide Threat Assessment (DNI 2016).

5. Force combinations of variables, elements, factors, technologies, which, when used in a different way, can generate harmful effects, exaggerating these capabilities. Making questions is needed: what if?, what effect? and what could happen next?

6. Identify the most relevant present and future actors exaggerating, varying or combining their objectives. Think not only of convergences of interests but also of differences. For example, so far we had not thought about a competitor to Al Qaeda. There are several analyses about possible links between terrorism and organised crime but not about rivalries and possible confrontations and clashes (as it happens with organised crime organisations). Do not forget that dystopias are always not only the result of a more or less uncontrolled environment that arises spontaneously, but also the product of human factor.

7. Consider the present and future battle of power: corporations, media, occult powers, bureaucracy, technology, ideas and religions.

8. Do not forget the past. The future has many elements from the past, though in a different political, social, economic, geographical or technological environment. Mankind has suffered quite dystopian events, such as Nazism, wars, massacres and genocides, pandemics and serious natural disasters. The need to know as deeply as possible the past and present does not mean we have to reflect it on the scenario that is to be created. It is one of the main cognitive challenges of this phase.

    Methodological note: This challenge is latent in the dystopian literature or science fiction if we observe that it reflected the greatest fears of each moment of the history. During the nineteenth century, socioeconomic inequality, peace or war focussed the main dystopian narratives. Today, the focuses that guide new scenarios are the hegemony of technology, the loss of individuality or the collapse of democracy. Thus, although it is still plausible to build dystopias based on current fears, developing the ability to generate unknown fears helps to design facts far from reality.

Aldous Huxley is an example to follow in this regard. With his great work "Brave New World" he was able to develop a scenario where, away from the war of all against all or a global collapse, he created an idyllic and apparently happy world, but devoid of imagination, creativity and individual freedom.

9. These stories have very marked characters, protagonists who rebel, struggling to escape, to change the world, who question the social or political order, helping to recognise the negative aspects of the future world that is portrayed.
10. Consider some common elements in literary or cinematic dystopias:

    - There is a conflict or a serious threat to humanity.
    - There may be violence.
    - The use of propaganda to control society.
    - Restrictions on freedom, information or knowledge.
    - A superior concept of freedom.
    - Surveillance.
    - Fears of new or external threats.
    - Dehumanisation and loss of the natural state of the world.
    - Loss of individualism.
    - The technology generates harmful effects and escapes human control.
    - Suffering, misery and search for truth or happiness.

11. All these elements, treated with creativity techniques, that attempt to ward off our thinking of present knowledge and common experiences, are parts of the necessary ingredients to build the scenario.
12. Build the scenario. A dystopias acquires value when it is nourished by elements, details, context, actors, relationships, enmities, interests, false appearances and technology. It is the most complete representation possible to a future world. It should reflect thought, emotion, and feelings. The objective is not to generate a picture, but a story.
13. Some ideas to include in the scenario: How is the society organised? What laws and restrictions exist? What are the penalties? How do people live? Where do they live? How is the economic system? What languages do they speak? What races or social classes exist? How it has come to that point? What steps have been followed? In which moment things worsened? What effects were produced? Are there rebelling groups fighting against the established power? How they were forged? What is the real situation? Who are the most important leaders? and Who worships the people?
14. Dystopias should leave ample space to the reader, encourage imagination, allowing them to draw causal lines, discoursing on the variables involved and the behaviour of the actors.

Methodological note: when the person or team that will design the dystopia is not used to work using creative processes or elements it may be useful to train them, previously, with the creation of *ucronias*. It is understood by *ucronia* the representation of a parallel world that, from a particular moment in our past, the existence of a fact or set of them made change completely the course of

history. From a cognitive point of view, it could be easier to develop a scenario of these characteristics because it begins from a known context.
(b) Deconstruction phase

This phase is a team process, which can be combined with other creative techniques such as structured brainstorming, what if, backcasting, or starbursting, following these steps:

1. Identify key concepts, risks and threats presented in the dystopia.
2. Identify the causes that have generated this undesirable future society, trying to determine indicators that can act as an early warning in the present.
3. Identify "weak signals" or current trends, to point out the need to monitor them.
4. Find sources of information that can complete the knowledge about the highlights, to broaden the debate and reflection.
5. Determine indicators that can act as alerts.

5 Case Study: Detection and Analysis of Wild Cards and Dystopias. Links Between Organised Crime and Terrorism

5.1 Presentation of the Use Case

This use case was developed by the authors of this chapter during a workshop of an EU-financed project, ePOOLICE (Early Pursuit Against Organised Crime Using Environmental Scanning, the Law and Intelligence Systems https://www.epoolice.eu) in Las Palmas, Spain (March 2015), with the presence of project partners and representatives of European law enforcement agencies. Previously, attendees were asked to read:

- The future of counter-terrorism in Europe. The need to be lost in the correct direction. Blanco and Cohen 2014. *European Journal of Future Research.* Springer.
- Exploring tomorrows organised crime. *Europol, 2015* (Europol 2015).
- To what extent do global terrorism and organized criminality converge? General parameters and critical scenarios. *Luis de la Corte Ibáñez, 2013, pp. 353–380.* (de la Corte Ibáñez 2013).
- Why Organized Crime and Terror Groups Are Converging. *Steven D'Alfonso, 2014* (D'Alfonso 2014).

The trainers presented 18 short stories, which established possible future relationships between organised crime and terrorism.

Applying a "what if" methodology, participants should identify trends, drivers, indicators and outcomes, following the methodology of trend analysis explained in Chapter "Macro Environmental Factors Driving Organised Crime" of this book.

The stories presented possible or probable scenarios in future (based on "scenarios design", in fiction films or books, or future essays—for example, "The digital future", Schmidt and Cohen (2014), leaders of Google and Google Ideas).

Participants, divided in groups, wrote their observations using this model:

TRENDS	DRIVERS	INDICATORS	OUTCOMES

Once we had identified the common trends and drivers, participants applied a method for evaluating the importance of each one (impacts and probability), in order to have a common priority list, and trying to build a system for monitoring their evolution. The model applied was as follows (Fig. 1).

5.2 Selection of Fictional Stories Proposed

1. 2032. Governments are controlling information, minds and sentiments of the society around the whole world. Governments are developing programs for "mental hacking". Lots of people, called "Free Society", hide from the government surveillance living underground, and using new parallel encryption systems to avoid being detected. They attack, using "guerrilla" and "lone wolf"

Assessing Trend Relevance

		Consider trend impact globally					Consider trend impact on your organisation			
Timeframe		**Scope**		**Impact**		**Likelihood**		**Urgency**		
When will trend begin to have an impact?		What is likely future uptake of this trend?		What is likely future impact of this trend?		What is the likelihood of the trend having an impact on your organisation?		How quickly does your organisation need to respond to this trend?		
Assessment	Rating	Assessment	Rating	Assessment	Rating	Assessment	Rating	Assessment	Rating	
1-4 years	5	Global	5	Significant	5	Almost Certain	5	Now	5	
5-9 years	4	Widespread	4	Major	4	Likely	4	Within 3-5 years	4	
10-14 years	3	Niche sector/market	3	Moderate	3	Possible	3	6-9 years	3	
15-20 years	2	Organisations	2	Minor	2	Unlikely	2	10-15 years	2	
20+years	1	Individuals	1	Insignificant	1	Rare	1	16-20 years	1	
Never *	0	Non-existing*	0					20+ years**	0	

*Before you assign "Never" or "Non-existing" to a trend, make sure you have tested your assumptions, and identified your blind spots. Ask what would have to happen to make the trend a reality? Only then should you feel comfortable assigning these categories to a trend.
** Even though the urgency to address these trends is long-term, consider keeping them on your scanning 'watch list'.

Assessment Total	Decision	Comment
	What might you do now?	
Between 20-25	Act now	You need to make a decision now about whether or not your organisation needs to respond to this trend. Consider how to respond and include in your current strategic plan if appropriate. If you decide not to include in your plan, then add to your watch list.
Between 15-19	Manage	You need to consider now how you might respond to these trends as they continue to emerge. It would be a good idea to include actions in your plan that allow you to act quickly if you need to.
14 and under	Watch	These trends are unlikely to have an impact on your planning in the medium term. To prevent future surprises, keep these trends on your scanning watch list.

Fig. 1 Assessing trend-relevant framework. Adapted from thinking futures for shaping tomorrows http://thinkingfutures.net/wp-content/uploads/Assessing-Trend-Relevance-Feb-10.pdf

models, big data centres of the government. They finance their operations trafficking with natural products, because all the products that are sold in legal markets are transgenic.
2. 2035. Several years later, rebel groups (now bigger and stronger) have developed their own drones, beginning a war between drones from the government and these ones. Every open public space represents a big security threat. "Free Society" is considered an international terrorist group, although they have a great and growing public support of their cause.
3. 2016. Dawn-esh, a terrorist group, desires to send foreign fighters to Europe, especially to Italy. For this purpose, they plan to use classical illicit trafficking routes with the support of several corrupt security borders agents. At the beginning, they started their contacts with organised crime in Libya. In order to use their routes they signed a clear pact: organised crime shares its routes if Dawn-esh shares its war tactics and, in some cases, if they do their "dirty work" for them.
4. 2021. A cyber attack in several European countries is affecting air traffic control and electrical systems. This action is followed by a terrorist attack in the streets, as it happened in Mumbai in 2008, when 10 members or Lashkar-e-Taiba carried out a series of 12 coordinated shooting and bombing attacks. Explosions occur in major European cities. The chaos produces a lack of energy supply, food and water.
5. 2017. The Colombian government signs a final agreement with the FARC. This new situation produced a displacement of activities, from part of its members, towards organised crime. As the control of territories in Colombia is assured by the new "Gendarmerie", they look for other criminal markets, being Europe and Spain a good opportunity for trafficking activities as it happened previously with Mexico. In the most attractive regions to settle (mainly in the Mediterranean coast and Galicia) they have to deal with the presence of anarchist organised motor gangs who have these markets under its control.
6. 2030. Smart cities are the new urban rule. 60% of the world population lives in there. Uncontrolled urbanisation has generated negative effects, highlighting the social polarisation and the appearance of urban ghettos. Many new urban tribes are emerging: antisystem movements, far-right groups and small local cells of the group "Humanist" that are violently repressing any sign of religious behaviour.

Clashes between these groups are common, and they are conquering several geographical areas. In the best case, these groups are holding talks with local authorities, never with national institutions. They do not recognise their authority. The territory outside government control already exceeds the 20% of the surface.
7. 2028. New commercial and routes are implemented, especially through Antarctica. Taking advantage of this situation, organised crime groups are signing agreements with insurgent groups to attack maritime border-crossings (such as "el Estrecho"). Trade and energy markets are being greatly affected by the attacks.

8. 2023. Far-right groups, from Eastern Europe, are manufacturing weapons with 3D printers. A new terrorist organisation from Kazakhstan (former members of the old Al Qaeda) has displaced their activities to Romania, Moldova and Ukraine, winning the control of the regions and controlling the traditional markets of organised crime groups.
9. 2020. Dawn-esh has attacked a shipment transporting drugs for the Yakuza, between Japan and India. The losses have triggered a war between organised crime and terrorist groups. The Italian mafia finally gives its support to the Yakuza. In retaliation to this alliance, lone terrorists ("lone wolves") begin to commit attacks against members of the Mafia in European countries.
10. 2022. Boko Haram (BH) has taken over Al Qaeda in the Islamic Maghreb and has extended its presence to the south of Algeria. They desire to attack Europe. For this purpose, BH has sent several shipments of smuggling gin that will be sold in southern Europe. They have only poisoned some of the bottles, making difficult to detect the origin of the attack. Alcohol smuggling is a major challenge for Europe after recent tax increases to these products as a way to tackle the public deficit. More than 200 people are killed.
11. 2017. European authorities have detected the sale of highly enriched uranium (HEU) from organised crime groups from Eastern Europe to communist groups with international presence, which are combating in Ukraine, Kurdistan, Iraq and Yemen. It is impossible to detect the origin of the product and the quantity stolen, but it increases the risk of use of a dirty bomb by non-state organisations.
12. 2018. There is a massive leak of sensitive information about security, affecting national armies and law enforcement agencies, with personal information of their members. A kidnapping and beheading international campaign led by a terrorist group begins, with several attacks from "lone wolves" and "social wolves" (self-radicalised individuals over the Internet) in New York, London and Paris. Members of the same families, especially brothers, and fathers and sons, commit some of the attacks.
13. 2019. Several insurgent and terrorist groups were financed by states, but this support is more difficult now. So, these groups need to diversify their incomes. They attack the vulnerabilities of virtual currencies, i.e. Bitcoin, and benefit from several activities in the Dark Web, where criminal organisations sell the "pack of the lone wolf" in a web called "Heaven Road": 3D guns, explosive belts (optional), instructions to make a bomb in your kitchen (or the kitchen of your mother), a hair ribbon, the new version of "A new religion for Dummies 2.0" and a list of the members of army and LEAs leaked by other groups.

For training purposes, they use programs based on virtual and augmented reality: videos, games and widgets. The consumers of these products feel the "emotion" to contribute to what they consider a better future. These videos achieve a more effective communication and incorporate more information in different formats (video, text, image, smell, emotion and feelings). They are perfect for indoctrination, training (no need to travel to classic training camps) planning and simulation of attacks.

14. 2032. In India, robots are being prepared for terrorist attacks, with advanced IA systems. Hackers that are working for the group "Free Society", which financed their activity trafficking with humans, not for exploitation, but to put them safe from autocrat governments and to help them abandon conflict countries, were previously developing these robots.
15. 2018. A new global organised crime group, "MafIA", with huge technological resources and advanced Intelligence Artificial systems, wants to control the world. Its leader wants to copy the jihadi style of the "lone wolves" and, with developed systems of communication, using the power of images and videos that show the impact of their criminal activities, and social media, is calling to unemployed youth all over the world to join this group and to act as "lone criminal wolves", facilitating them the resources to commit their crimes.
16. 2021. A far-right group attacks the big data centres of the UK government. Extracting entities and analysing information, they identify Jews living in the country and access to their medical records. They try to assassin them using cyber attacks and provoking heart attacks.
17. 2023. The world suffers an extreme shortage of resources. Access to Internet is arriving to places with historical local and regional conflicts. This facilitates the emergence of new conflicts, often based in the extraction, production or management of resources such as water, minerals, foods, raw materials or sources of energy. The main way of attacking their enemies is through the use of improvised explosive devices (IED), which are being incorporated into drones manufactured by "MafIA", an organised group which controls the "Crime as a Service" market.
18. 2019. A new modus operandi in kidnappings is producing new concerns. Not persons, but data are being kidnapped, and criminal, insurgent and terrorist groups are introducing this new modus operandi: "data for ransom". In case governments, private enterprises or individuals do not pay the ransom, they threaten to "behead" the data (gradually eliminating them).

5.3 Conclusions of the Case Study

The exercise allowed the identification of 35 common trends affecting the development of terrorism and organised crime, and the links between both threats.

Political trends
1. Globalisation of crime and terrorism.
2. Conflicts, causes, effects and shared territories.
3. State fragility.
4. Corruption.
5. Growing surveillance.

6. Increase power of non-state actors.
7. Growing extremism, populism and nationalism.
8. International military interventions, post-conflict management and democracy process.

Economic trends

9. Inequality.
10. Economic and criminal economic globalisation. "Crime as a Service".
11. Unemployment.
12. Economic and financial crisis.
13. Tax havens and lack of financial controls.
14. New commercial routes.
15. New criminal and terrorist financing activities.
16. New ways to transfer money (virtual currencies).
17. Shared territories of production or transport of drugs.

Social trends

18. Demographic changes (ageing, migrations and decrease in working population).
19. Migrations.
20. Urbanisation.
21. Growing individual empowerment.
22. Growing social conflicts.
23. Evolution of Western values.
24. Opportunities for common social support to terrorism and organised crime.
25. Shared places: prisons, borders, ghettos…

Technological trends

26. IT development (Internet, social media, mobile devices).
27. Big data (cloud computing, privacy, cyberattacks, IoT).
28. Transportation and logistics.
29. Technologies with double use (3d printings, drones, IED…).
30. Smart cities.
31. Nanotechnology and robotics.

Environmental trends

32. Resources scarcity.
33. Climate change.
34. Pandemics.
35. Natural catastrophes.

These variables could provide input to establish a monitoring system, with drivers and indicators, in order to evaluate the evolution of links between organised crime groups and terrorist groups and individuals. This approach has been part of research European programmes, for example, ePOOLICE, or the COMPSTAT system used by the New York Police Department.

6 Conclusions

We cannot predict the future. The large number of variables involved in organised crime, the very rapid change in our societies and the complexity of the interrelations between actors make very difficult to determine the future of the phenomenon. Long term is now very short.

Given this situation, it is worth thinking about future? Emphatically, yes. From the moment we think in future, we introduce causal elements in its development. Thinking about the future is the first step in its construction.

There are two ways to approach the study of the future: the first, from the present, through a descriptive analysis and detection and monitoring of new trends; the second, from the itself, in a backcasting process, trying to avoid the biases that our knowledge and our previous experiences produce. This requires a high exercise of imagination and creativity.

Nicholas Taleb affirms that we will be unable to detect black swans, because subsequent analysis will overestimate our capabilities for prediction and underestimate the importance and "role of chance in life and in the markets". That is definitely true. Our ability to discover the so-called unknown unknowns (Rumsfeld) is very limited. But it is possible to work in the "unknown knowns" (that is to say that we could know several things if we asked to the right person or if we used an appropriate methodology) and "known unknowns" fields (that is to say there are things that we now know we do not know, so we would need to improve our research efforts). The past and the present, different geographical environments, and current global trends can combine different elements and generate alternative scenarios, some of them more favourable than others. We must avoid from the present worst scenarios. It is an urgent and inescapable task. And we must do our best to get the most positive scenarios. The freedom of our societies depends on it, tackling criminal phenomena that pose serious attacks on individual and collective rights and freedoms, human rights, good governance and democracy.

References

Berger, G. (1964). *Phénoménologie du temps et prospective*. Paris: Presses universitaires de France.
Berger, G. (1967). *Étapes de la prospective*. Paris: PUF.

Blanco, J. M. (2015). Gestión del conocimiento y cultura de seguridad. *Cuadernos de la Guardia Civil*. N° 50 especial. Madrid: Guardia Civil
Blanco, J. M., & Cohen, J. (2014). The future of counter-terrorism in Europe. The need to be lost in the correct direction. *European Journal of Foresight Research, 2*(1), Springer. Retrieved May 10, 2016 from http://link.springer.com/article/10.1007%2Fs40309-014-0050-9
Blanco, J. M., & Jaime, O. (2014). Toma de decisiones y vision de futuro para la seguridad nacional. en De la Corte, L. y Blanco, J. M. (Eds.), *Seguridad Nacional, amenazas y respuestas*. Madrid: LID.
Clark, R. M. (2013). *Intelligence analysis. A target-centric approach*. London: CQ Press.
D'Alfonso, S. (2014). Why organized crime and groups are converging. Retrieved May 10, 2016 from https://securityintelligence.com/why-organized-crime-and-terror-groups-are-converging/
de la Corte, L. (2013). To what extent do global terrorism and organized criminality converge? General parameters and critical scenarios. *Journal of the Spanish Institute of Strategic Studies*. Retrieved May 10, 2016 from http://www.ict.org.il/Article/997/To%20What%20Extent%20Do%20Global%20Terrorism%20And%20Organised%20Criminality%20Converge
ESPAS. (2015). *Global trends to 2030: Can the EU meet the challenges ahead?* Retrieved May 9, 2016 from http://europa.eu/espas/pdf/espas-report-2015.pdf
Esposito, E. (2011). *The future of futures*. Cheltenham (UK): Edward Elgar Publishing Limited.
Europol. (2015). Exploring tomorrow's organised crime. Retrieved May 10, 2016 from https://www.europol.europa.eu/publications-documents/exploring-tomorrow%E2%80%99s-organised-crime
Godet, M., & Durance, P. (2011). *La prospectiva estratégica para la empresa y los territorios*. París: DUNOD y UNESCO.
Goodman, M. (2015). *Future crimes. A journey to the dark side of technology—and how to survive it*. London: Penguin Random House
Gordon, T. J. (2016). *Anti-terrorism scenarios. international study on counterterrorism, millennium project*. Retrieved May 10, 2016 from http://www.millennium-project.org/millennium/antiterrorism.html
Heuer, R. J., & Pherson, R. H. (2014). *Structured analytic techniques for intelligence analysis*. CQ Press
Institute for the Future. (2008). *Knowledge tools of the future*. Retrieved May 10, 2016 from http://www.iftf.org/uploads/media/SR-1179_FutKnow.pdf
Lowenthal, M. M. (2013). A disputation on intelligence reform and analysis: My 18 Theses. *International Journal of Intelligence and Counterintelligence, 26*, 31–37.
National Commission on Terrorist Attacks. (2004). *The 9/11 Commission Report*. New York: Norton.
National Intelligence Council. (2016). *Global trends 2030. Alternative worlds*. Retrieved May 10, 2016 from http://globaltrends2030.files.wordpress.com/2012/11/global-trends-2030-november2012.pdf
Petersen, J. (1997). *Out of the blue how to anticipate big future surprises. The Arlington institute* (2nd ed.). Lanham: Madison Books.
Rand Corporation (2013). *Predictive policing. The role of crime forecasting in law enforcement operations*. Retrieved May 10, 2016 from http://www.rand.org/content/dam/rand/pubs/research_reports/RR200/RR233/RAND_RR233.pdf
Schmidt, E., & Cohen, J. (2014). *El futuro digital*. Madrid: Ediciones Anaya Multimedia.
Schwartz, P. (1996). *The art of the long view*. New York: Currency Double day.
Svendsen, A. M. D. (2012a). *The professionalization of intelligence cooperation: fashioning method out of Mayhem*. Basingstoke: Palgrave Macmillian.
Svendsen, A. M. D. (2012b). *Understanding the globalization of intelligence*. Basingstoke: Palgrave Macmillian.
Taleb, N. N. (2012). *Antifragile: Things that gain from disorder*. Random House.
The Chapman University. (2015). *America's top fears 2015*. Retrieved May 10, 2016 from https://blogs.chapman.edu/wilkinson/2015/10/13/americas-top-fears-2015/

The Millennium Project. (2009). *Futures research methodology Version 3.0* (C. G. Jeroneme, T. J. Gordon, Eds.).
Toffler, A. (1971). *El "shock" del Futuro*. Barcelona: Plaza y Janes.
U.S. Government. (2014). *Big data: Seizing opportunities, preserving values*. Retrieved May 9, 2016 from https://www.whitehouse.gov/sites/default/files/docs/big_data_privacy_report_may_1_2014.pdf
U.S. Government. (2016). *Worldwide threat assessment of the U.S. Intelligence*. Retrieved May 10, 2016 from http://www.intelligence.senate.gov/sites/default/files/wwt2016.pdf
World Economic Forum and Boston Consulting Group. (2016). *New vision for education: Fostering social and emotional learning through technology*. Retrieved May 9, 2016 from http://www3.weforum.org/docs/WEF_New_Vision_for_Education.pdf
World Economic Forum and Boston Consulting Group. (2015). *New vision for education: Unlocking the potential of technology*. Retrieved May 9, 2016 from http://www3.weforum.org/docs/WEF_New_Vision_for_Education.pdf
World Economic Forum. (2016). *The global risks report 2016*. Retrieved May 10, 2016 from http://www3.weforum.org/docs/Media/TheGlobalRisksReport2016.pdf

Part III
Use-Cases

Operation Golf: A Human Trafficking Case Study

Bernie Gravett

Prologue
By the editors

How Such Investigations Can Benefit from Integrated Intelligence-Led Policing

One aspect that expert intelligence analysts often emphasise as one of the biggest challenges in this field is to address the gaps between the different types of intelligence. We highlight some of them:

- *Strategic intelligence.* Its aim is to support the decision-making process, understanding the phenomenon and generating knowledge to produce strategies, plans and actions. Most of the chapters in this book have addressed this dimension (environmental scanning, PESTLE, indicators, trends), including the so-called estimative or prospective intelligence, which tries to answer: what seems to be happening? what is really happening? what will happen in the future? and what should we do now?
- *Tactical intelligence.* Its aim is planning, so it needs more details: place, time, strengths, weakness, enemies or competitors, etc.
- *Operational intelligence.* Its aim is to collect and analyse information in order to conduct a campaign or operation.

Several times, strategic analysts are far from the information obtained by police operations taking place on the streets or in cyberspace and often perceive that security policies—and therefore the derivative actions—lack strategic and future

---

B. Gravett (✉)
Specialist Policing Consultancy Ltd., Oxfordshire OX25 3RH, UK
e-mail: Bernie.Gravett@specialist-policing.co.uk

vision. Moreover, it is also common that operational officers and analysts consider strategic analysts as theoretical or academic, without knowledge about what really happens on a criminal phenomenon.

Both have a point, but not entirely. We tend to analyse problems from very partial visions, our own expertise and professional performance, our knowledge and our previous experiences. Thus, there is much knowledge, but absolutely fragmented, which prevents having the Big Picture to analyse a criminal phenomenon.

Strategies are accurate to develop effective and efficient operational actions. They introduce the long-term vision, establish the framework for the development of plans and actions, and should evaluate the impacts. Strategies explain the causes, not only trying to influence the symptoms of the criminal phenomenon, but also supporting decision-makers in implementing comprehensive policies to act against criminal groups, against their actions, against the effects and impacts generated and, especially, against the causes, facilitators or enhancers of the criminal phenomena.

Operative police officers have the information on what is happening in the criminal landscape (actors, crimes, modus operandi, financing activities, firearms and weapons, hot spots), but do not have the time or methodologies to explain the causes of the phenomenon and its evolution, or possible measures to apply.

Thus, the strategic and operational knowledge should be a continuous line, working bidirectionally: operational data must reach the strategic line that supports decision-making; strategic knowledge produces plans and actions to enhance the effectiveness in the fight against organised crime.

A similar problem arises with the types of intelligence in terms of the sources used. Despite the current proliferation of denominations, we can highlight the use of open sources, human sources and technological sources. It is customary in the field of intelligence to avoid ongoing debates about which source is better. Any interpretation that try to put others ahead of any of the types of sources will be wrong. It will depend on what topic is being analysed and especially for what purpose to determine the appropriate sources. In addition, intelligence analysts should never waste the value of any source that can be useful, being the right solution to look for a combination of all types. Open sources have advantages such as ease of access and low cost, and great disadvantages: a huge amount of information and the need for validation. Human sources have better knowledge of criminal actors and activities but may have a tendency to lie. Although the great probative power of an image or video in investigations, sometimes technological sources have also shown their limitations (as in the case of the alleged weapons of mass destruction in Iraq).

Examples on How Open Source Intelligence Can Help Such Investigations

In the intelligence-led policing framework outlined above, open source intelligence could help investigations like Operation Golf in several ways, using methodologies and technologies like those explained in other chapters (including those developed in the ePOOLICE project).

Thus, environmental scanning and early warning could be used for early detection of emerging threats, i.e. before they unfold, and can provide useful knowledge on the environmental factors giving rise to a current threat or act of organised crime. Some examples of how an investigation like Operation Golf case could benefit from open source intelligence are:

1. *Indications of such organised crime*

Monitoring open data (environmental scanning) for indications of organised crime involving trafficking and exploiting children in some urban area:

- Incidents of crime typically done by children exploited by the organised crime networks, cf. the following case description (open data: news articles and crime statistics)
- Complaints by neighbours to houses or apartments that appear to be used in an abnormal way, housing many people, in particular children (open data: news articles and community statistics)
- Abnormal large or increasing number of shops of a particular kind (e.g. nail bars) that may be used as cover for trafficked humans and/or for money laundering (open data: news articles and business registers).

2. *PESTLE factors*

Monitoring and analysis of PESTLE (political, economic, social and demographic, technological, legal and environmental) crime relevant push and pull factors driving such organised crime. This is useful for analysing and understanding the background for the crime threat and hence for making better, knowledge-based decisions for preventing, investigating and mitigating the threat. (Open data: statistics such as Eurostat, reports, news.)

3. *Immigration statistics*

Monitoring immigration statistics for data that are consistent with expectations by PESTLE factor analysis and OC indications or may give rise to a targeted analysis.

Furthermore, the operational information in the case could be useful for strategic analysis based on open sources, allowing knowing actors involved, slang language, and weak signals about changes in criminal groups, nationalities involved, modus operandi, objectives, criminal places or firearms used. This information must be merged and integrated with the one provided by the open source monitoring systems.

1 Introduction
By Bernie Gravett

Operation Golf was a joint investigation between the Metropolitan Police Service (MPS), UK,[1] and the Romanian National Police (RNP) targeting a specific Romanian Roma organised crime group that is trafficking and exploiting children from the Romanian Roma community. This community is one of the poorest and most disadvantaged communities in Europe. This organised criminal network (OCN) has increased its activity since accession and is now trafficking entire families for forced criminality and benefit crime. The MPS operation was commissioned by Commander Steve Allen following a 786% increase in Romanian sanction detections in the first 3 months of 2007 across the MPS.

2 Background

In the summer of 2004, Chief Inspector Bernie Gravett and Inspector Colin Carswell, both officers from Westminster Police in central London, identified that Romanian Roma groups in central London were committing a considerable volume of level 1 crime.[2] The offences were being committed by children and young persons and included begging, shoplifting, distraction theft, deception, fraud, credit card cloning and robbery. Children as young as 6 years old were stealing handbags and mobile phones from coffee shop customers.

Intelligence checks revealed that the families of the children were illegally in the UK at this time, having been smuggled into the country in cars and lorries by the OCN.

This led to an operation conducted in cooperation with the UK Immigration Service and the removal from the UK of 407 individuals concerned in these crimes. This operation was named 'Golf' and was conducted from October 2004 to July 2006.

As a result, theft offences in Westminster were reduced by 29% and pick pocketing was reduced by 24%.

All recorded crime in Westminster dropped by 2%.

---

[1]The Metropolitan Police Service employs around 31,000 officers together with about 13,000 police staff and 2600 Police Community Support Officers (PCSOs) (Source: http://content.met.police.uk/Site/About, accessed 15 October 2016).

[2]Level 1 criminality is defined as 'Local Issues—usually the crimes, criminals and other problems affecting a basic command unit or small force area' (Source: https://www.lightbluetouchpaper.org/2006/02/06/mysterious-and-menacing/, accessed 15 October 2016).

> **CASE STUDY: Girl A DOB: 01/01/1986—now 24 years old**
> Girl A is one of 1087 children taken from Romania pre accession. She was driven out of Romania by the gang along with 5 other children. Her journey took her into Hungary and across Europe. She first came to notice in the UK in 2002 when she was 16 years old. She was arrested for theft within Westminster Borough. She received a juvenile reprimand for this offence. Since then she has acquired a total of 17 convictions and 3 cautions, with offences of shoplifting, distraction thefts and failing to answer court bail. She was arrested a further 6 times but the offences were not proceeded with. She has served a prison sentence in Holloway women's prison.
>
> She has total of 8 alias names and 9 dates of birth. There are 43 intelligence reports on her in London. She has been arrested predominately in Westminster but also Enfield, Camden, Hammersmith and Kensington. She is also known to commit offences in Surrey, City of London, and within the area covered by BTP.[3]
>
> In 2006 she was moved by the gang to Spain but returned to the UK in 2007 following accession of Romania into the EU.
>
> She has numerous associates all of whom have convictions on PNC[4] and are well known to Police within the Metropolitan Police District. She continues to live in poverty, gaining no benefit from the criminality.

A significant amount of learning and intelligence was gained through this operation, and it was identified that Romanian Roma Organised Crime Networks were behind these crimes and that they were using children as tools to commit the offences. An outcome of this was that Chief Inspector Gravett was seconded to the Foreign and Commonwealth Office and travelled to Romania to assist the Romanian National Police in capacity building in preparation for entry into the European Union. At the same time, Inspector Carswell joined SCD6[5] International Crime Coordination Unit with specific responsibility for Romania and other Balkan states and the related threats posed to London.

---

[3] The British Transport Police (BTP) is a special police force that polices railways and light-rail systems in England, Scotland and Wales, for which it has entered into an agreement to provide such services (Source: https://en.wikipedia.org/wiki/British_Transport_Police, accessed 19 October 2016).

[4] The Police National Computer (PNC) is a computer system used extensively by law enforcement organisations across the UK (Source: https://en.wikipedia.org/wiki/Police_National_Computer, accessed 16 October 2016).

[5] A unit of New Scotland Yard's Specialist Crime Directorate.

3 The First Indication of Trafficking

During this period, the officers became aware of Operation 'Girder'. This was an operation executed by the Serious Organised Crime Agency (SOCA)[6] where it had been identified that a Romanian OCN had trafficked 21 children into the UK for the purposes of committing volume crime. This operation led to 8 adults convicted for illegal facilitation of children into the UK.

Only 3 of these children were identified and recovered.

The gangs used a Czech Roma woman, Ana PUZOVA, to bring the children to the UK. At this time, Czech passports only listed the children of the family but there were no photos. Immigration intelligence identified that PUZOVA was making trips into the UK with children but leaving alone. She was stopped entering the UK with three children passing them off as her own. When questioned, it was identified that the children were Romanian Roma from Tandarei, Romania, and were Roma.

> Ana PUZOVA is a Czech national with 8 natural children of her own. The link was that she is Roma. She was paid 1000 Euro per trip. At this time, the gang charged the Roma families 1000 Euro per child for them to be trafficked. The children were taken to Italy and Spain by the gang and from there flown with Ana into Stanstead and Luton airports. The children were then passed back to the gang and distributed across the UK.
>
> Puzova, who was pregnant with her ninth child at the time of her arrest, had come to police attention two months earlier after flying into Luton from Barcelona with two children. Further investigations revealed a pattern of frequent travel between British airports and destinations in Spain and Italy.
>
> She pleaded guilty at Chelmsford Crown Court in 2006 to six charges of facilitating the unlawful entry of children into Britain and was jailed for three years.

The gang's leader for this part of their operations was identified as Remus KVEC. While KVEC was based in the North West of England, there were links to addresses in North East London. He was sentenced to 8-year imprisonment for running the trafficking ring. However, this was the tip of the iceberg.

Operation Girder led the Romanian National Police (RNP) to open an investigation into the Romanian OCN involved. This identified that the OCN had trafficked 1087 identified children out of Romania. The evidence is that the majority of these children have been, or were being, exploited by being made to beg and commit theft in a number of European countries. The OCN and children all

---

[6]The Serious Organised Crime Agency (SOCA) was a non-departmental public body of the Government of the UK which existed from 1 April 2006 until 7 October 2013 (Source: https://en.wikipedia.org/wiki/Serious_Organised_Crime_Agency, accessed 16 October 2016).

originate from the single town of Tandarei in South East Romania. Both the victims and the OCN are from within the Romanian Roma community.

> The RNP investigation have identified the trafficking routes and method used and that the OCN is operating across Europe primarily in UK, Italy, Spain and France. However, their challenge was that the exploitation was taking place outside Romania, and what they see were the gangs getting richer on the proceeds. The most visible aspects are the building of large houses, the purchase of expensive vehicles and the possession of large amounts of disposable cash.

In January 2007, Romania joined the EU. Within 3 months, crime in London committed by Romanian nationals went up 786%.[7] Analysis showed the offences to be predominantly theft committed by children from within the Romanian Roma community. The London Borough of Westminster was particularly affected by this rise in crime. This growth was caused by the gangs moving people into the UK from Spain and Italy, now much easier as the border checks were removed.

In April 2007, Commander Steve Allen, BOCU[8] Commander for Westminster, commissioned a small team, led by the now promoted Superintendent Gravett and Chief Inspector Carswell, to examine the causes behind the rise in crime and the links to Organised Crime. This was again given the name 'Operation Golf' as it was already known that this was the same OCN.

Research on criminal records showed that 200 of the 1087 victims identified by the RNP were criminally active in London in the summer of 2007 and had convictions in 32 other Police Force areas.

The OCNs concerned are involved in the large-scale trafficking of children and adults who are forced into criminality to pay off debts bearing extortionate interest rates. OCNs abuse cultural rules whereby if a person defaults on a debt, they become the slave of the debtor, as does the belongings and property. This includes the family of the defaulter.

To test the finding of the research, the UK team set up Operation Caddy.[9] This focussed on the town of Slough to the west of London. Each day, up to 50 Romanian Roma would travel by train to central London. From here, they would split up and move across London committing crime focussing their activity on crowded places.

---

[7]PIB Nationality Index reports 2007.

[8]Borough Operational Command Unit.

[9]It is common in British policing operations to run smaller sub-operations to action specific lines of enquiries and to manage the resources required. These sub-operations will be given their own name for management purposes.

On the 28 January 2008, the team executed search warrants at 16 addresses in Slough. This resulted in the arrest of 34 people for a variety of crimes with over 200 items of stolen property recovered.

The most important aspect was that within the 16 small terraced 3-bedroom houses, police found 211 people; the majority of which were children. Ten children we recovered when it was found that their parents were not present.

Some houses were occupied by three families with children sleeping on the floor on sheets, and in one case a child had her bed in the bath. Each house was in squalor with little or no food present. The operation was conducted with the support of local social services. We were shocked to discover 60 children under the age of 10 that local social services had no knowledge of. Only three children were attending school out of 107 discovered, and these were the sons of the gang leader in the town. No girls were in education.

The arrests were for:

- Trafficking children into the UK
- Trafficking children around the UK
- Child neglect
- Theft
- Handling stolen goods
- Money laundering
- Benefit fraud
- Being wanted for failing to appear at court.

In addition, there were substantial cash seizures and a large amount of documentation found (Fig. 1).

Operation Caddy analysis
16 addresses 211 people encountered
103 adults 60% with criminal records
33 Juveniles 78% criminal records
74 minors (under 10) 47% on MPS intelligence for committing crime in London

Prevalence in under age pregnancy, some as young as 13 years old
Only 3 children in education
60 minors not known to Slough Borough Council
54% reduction in pick pocket offences in Westminster for the following 6 months

Ten children were treated as potential victims of trafficking and placed into emergency police protection and passed into the care of social services.

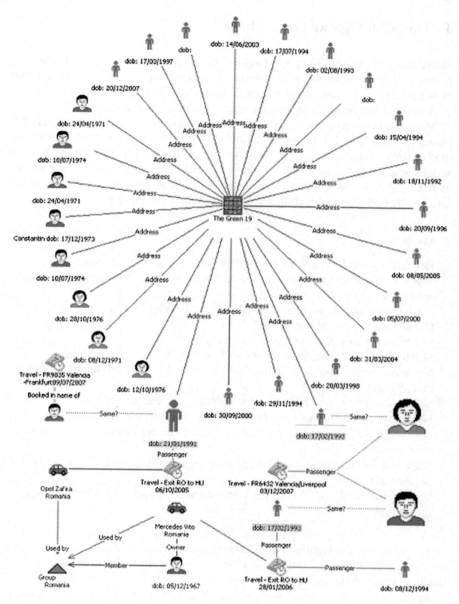

Fig. 1 One house with three families but also three unaccompanied minors (Sanitised (i.e. sensitive and classified information has been removed.) mapping chart of one element of the OCG)

The days following, the operation saw the parents of these children arriving from Romania and Spain with a variety of accounts as to how their children were left with families that exploit children. Nine of the children were returned to their parents with care procedures placed around them.

4 One Child Spoke Out!: Maria

Maria was a 13-year-old girl from Tandarei. Both her and her sister were taken from Romania to the UK by the gang. Maria was placed with a family in Slough and exploited. Her sister has yet to be found, and it is assumed that she is also being exploited but has been given an alias to prevent discovery and recovery.

Maria initially gave a true account of her treatment and abuse. However, following the arrest of her father for trafficking her and telephone contact with her mother back in Romania, she retracted her story. This could potentially have had a serious impact at the trial but the diligent recording and gathering of evidence allowed police to 'prove' her first account. This led to the first convictions in the UK for the trafficking of a child for forced criminal exploitation.

> **Maria's story**
>
> Her father paid 200 Euro to the OCN to have her trafficked to the UK
> She was flown to Stanstead by Busioc Vasile with the flights pain for on a corrupted USA credit card
> Placed with an OCN family in Slough and controlled and exploited by Claudia Stoica and Marin Vasile.
> She was told to call them Uncle and Auntie
> She became the house slave in domestic servitude and forced labour
> She was driven to Surrey each day and left for 12 h to beg, sell the 'Big Issue' illegally and steal.
> She kept nothing, was beaten and searched at the end of her day.
> Her father cloned Maria's identity to exploit children in Valencia Spain
> 4 people were convicted for her trafficking into and around the UK for forced exploitation
> They were sentenced to a total of 24-year imprisonment for trafficking, child neglect and perjury.
> The urgent need for a JIT (joint investigation team) with Romania was recognised

This investigation highlighted the complicity of parents in the trafficking of their own children. While debt slavery is one aspect of how the gang controls the families, greed also plays a part.

In addition to the criminal prosecution, there was a parallel care case running concerning Maria's welfare. This was taken to the High Court with the outcome being that UK courts have to rely on Brussels'II decision that a child must be returned to their country of origin for the authorities to manage her welfare there. Subsequently, Maria was repatriated to Romania and she passed into the care of Romanian Social Services. She was later reunited with her mother. While she has not been re-trafficked she was later sold into marriage and became pregnant at age 14.

> **Brussels II—Jurisdiction in relation to parental responsibility**
> Article 66 applies in children's cases. This is the article that relates to member states where there are two or more systems of law. Any reference to habitual residence in the member state 'shall refer to habitual residence in a territorial unit'. This implies that jurisdiction lies with the courts of the territorial unit in which the child is habitually resident. Such an interpretation would be consistent with the provisions relating to divorce. On this view, Brussels II governs the distribution of cases within the UK, as well as distribution between EU member states.
>
> As a victim of trafficking who had only been in the UK for a few months, the court decided that Maria was 'habitually resident' in Romania and that Romania had jurisdiction in matters of her welfare. This applied despite police presenting a case that she would be at risk of retribution, harm and further exploitation.
>
> The court's decision was that it had to abide by Brussels II and Maria's safety and welfare was a matter for Romania.

This operation is unprecedented in identifying the sustained demands made on the Police and partner agencies who regularly encounter the victims through arrests and street engagement across the MPS. The operation has been extremely successful and is active in thwarting the OCNs; however, we are seeking support to enable us to continue with the long-term investigative commitment that will be required to bring the accused to trial and dismantle the OCN both here and in Romania.

> *It is a sad fact that children are bought and sold around the world, trafficked into and around the UK for the profit of others. It is a complex but hidden crime that is largely unseen by broader society and unrecognised by frontline police and social services.*

5 The Formation of the JIT, 1 September 2008

The reason for forming a JIT was due to the international nature of the OCN's operations and the fact that the crime and exploitation is occurring in the UK but the profits were being realised in Romania. This was the first EU JIT tackling human trafficking. In addition, the team were the first UK police force to set up and conduct a JIT with another EU state.[10]

---

[10] The only other case of a UK JIT was between the UK NCIS and the Dutch police in 2006; it lasted only 3 months and targeted drug trafficking.

The JIT was 70% funded by a grant successfully obtained from the European Commission. The MPS Territorial Policing Command covered the remaining salary costs of the MPS staff on the team.

> **What is a JIT?**
> Article 13 of the European Convention on Mutual Legal Assistance in Criminal Matters of 29 May 2000 and/or of the Council Framework Decision of 13 June 2002 on joint investigation teams (JITs) provides the legal basis of the arrangements for the conduct of JITs in EU member states.

The EU acknowledges that member states have different legal systems; however, it is possible to set up a JITs across the EU in accordance with the Framework Decision. The UK is a common-law jurisdiction, with a permissive legal system, i.e. anything can be done unless it is specifically prohibited or regulated to be done in a certain way. The government did not think it was necessary to introduce new legislation in order to substantially implement the Framework Decision, although minor legislative amendments were needed to implement a few parts of it (see paragraph below). So investigators and prosecutors can enter into a JIT, relying on the Framework Decision and/or the 2000 MLA Convention.

The UK introduced provisions concerning JITs are: Sections 103 and 104 of the Police Reform Act 2002, which relate to obligations in Article 2 and 3 of the Framework Decision, dealing with civil and criminal liabilities concerning members of JITs; and Section 16 of the Crime (International Co-operation) Act 2003, which implemented Article 1(7) of the Framework Decision.

The UK ratification was notified on 22.09.2005. It entered into force on 21.12.2005. The 2000 MLA Convention has not been directly incorporated into the law of the UK. Joint investigation teams, led by police officers and operating in the UK, must comply with the provisions on liabilities in Sections 103 and 104 of the *Police Reform Act 2002* and would be able to take advantage of the provisions involving dispensation with letters of request in Sections 16, 18 and 27 of the *Crime (International Co-operation) Act 2003*. Home Office Circulars 53/2002 and 26/2004 draw attention to a range of operational matters, such as the need under the Framework Decision to ensure that the team leader is provided by the competent authorities of the Member State in which the team is operating. In practice these arrangements ensure compliance with Article 1.

The further Home Office Circular 26/2004 issued on 26 April 2004 draws attention to the new legislation in Sections 16, 18 and 27 of the *Crime (International Co-operation) Act 2003*, which implements Article 1, paragraph 7 of the Framework Decision. This legislation enables police officers and customs officers who are members of a joint investigation team to apply for a search warrant and/or a production order in relation to criminal conduct abroad without a letter of request (Commission Rogatoire). The legislation for Scotland (Section 18) is similar, but not identical, to the legislation for England and Wales.

The authorities that can authorise a JIT are the investigation agencies (e.g. the Serious and Organized Crime Agency, the police and customs) and prosecution authorities (e.g. the Crown Prosecution Service, Revenue and Customs Prosecution Office and the Serious Fraud Office). It is anticipated that in most instances a decision will be made jointly by the investigation agency and the prosecution agency.

Home Office Circular (HOC) 53/2002 states that JITs may be established under the Framework Decision (FD) by the competent authorities in England and Wales, Scotland and Northern Ireland.

Operation Golf was a joint investigation team (JIT) between the MPS and the Romanian National Police. The full JIT partnership is Operation Golf (MPS), Romanian National Police, D.I.I.C.O.T. (Romanian Prosecutors Office), the United Kingdom Human Trafficking Centre (UKHTC), Crown prosecution Service, Europol and Eurojust.

The Strategic objectives of the JIT were to successfully:

- Investigate and prosecute OCN members both in the UK and Romania
- Disrupt their activities
- Identify, restrain and confiscate criminal assets
- Reduce criminality
- Minimise the exploitation of victims
- Improve victim identification and response to child trafficking by police and partners.

6 Arrests and Prosecutions

The following is a précis of information relating to all arrests of individuals of interest to Operation Golf, both known members of the Organised Criminal Network or their associates.

Since the inception of the JIT and late 2010, there were a total of 120 persons arrested in the UK linked to the OCN. This includes the first ever convictions in the UK for trafficking a child into the UK and only the second convictions for internal trafficking within the UK.

A summary of arrest and prosecution activity is as follows:
Arrests and prosecutions have been for:

- Human Trafficking Sec 4(1) Immigration Act (Into UK)
- Human Trafficking Sec 4(2) Immigration Act (Within the UK)
- Money laundering
- Obtaining Benefit by Deception Sec 106 Immigration Act
- Forgery and Theft Act offences.

In addition, the team Operation Golf has to date been accredited with three Organised Criminal Network (OCN) disruptions.

7 Support to Romanian Investigation

As stated above, the Op Golf investigation was a joint one with the Romanian National Police. This was due to the modus operandi (M.O.) of the gang where the children and families are exploited in the UK and the OCN hierarchy remains in Romania realising all the proceeds of the gang's activity.

The UK Op Golf team supplied the Romanian Team with a full and extensive evidential package to prove the exploitation of the children and families in the UK. This has included, in an evidential format, the following:

- Full list of criminal convictions and circumstances for the 168 children identified as being active in the UK.
- Full details, including statements, of all contact by Police with the 168 children and families (stops, verbal warnings, intelligence reports, etc.).
- Evidential product and statements surrounding money transfers via UK-Romania money service bureaus (MSB) for the principle suspects.
- Evidenced details of all appropriate adults appearing for the children who have been arrested for crime in London.

This substantial piece of work has now directly resulted in the Romanian authorities preparing to arrest and charge Romanian nationals, all part of the gang, with trafficking children to the UK.

> **Operation 'Longship' a test of the JIT framework**
> A significant problem for Romania is that the exploited children were in other jurisdictions. Because of the JIT, we were able to deal with this issue by flying the Romanian investigation team to the UK for Operation Longship. The UK team identified and recovered 27 children and provided them to the Romanian team for a 'Witness Hearing' under Romanian law on UK soil. A challenge for the team was that in Romania child witness must be represented by a Romanian lawyer. To deal with this, the Romanian party included 4 independent Romanian lawyers to oversee the process and ensure the children's rights were upheld. This was the first such action of its kind in JIT history.

8 Operation EUROPA Arrest Phase

The first phase of the Romanian arrest operation took place on 8 April 2010. This involved the execution of search warrants at 34 addresses in Tandarei and the arrest of 18 persons for trafficking and money laundering.

The Romanian operation involving over 400 officers was supported by 26 members of the Metropolitan Police whose roles included command, intelligence and 11 investigation teams to accompany RNP officers on the searches. Under the JIT agreement, the Metropolitan Officers were able to use their UK powers on Romanian soil. They were allowed to search for evidence and question suspects as part of the UK investigation.

This was the first time this had occurred in EU history and was a significant step in extending powers to combat international organised crime.

In addition to the arrests, the Romanian authorities seized 4 AK47 rifles, 12 hunting rifles, 12 shotguns, including military grade weapons, and 6 semi-automatic handguns. Other seized items included 25,000 Euros, £25,000 and 40,000 Romanian Lei, 13 high-value cars, 6 houses and a substantial amount of evidence linking the gang to the UK and other EU countries. One remarkable find was 10.5 kilos of gold, which is worth approximately €500,000. The proceeds of crime in the UK are often converted into cash and then into gold for couriers to take back to Romania. Gold is easy to conceal, and the story given is often that it was old family jewellery melted down. This is hard to challenge by the authorities.

The Romanian authorities charged 26 gang members with the following crimes[11]:

- Trafficking 181 children to the UK for forced criminality;
- Money laundering;
- Being members of an organised criminal network;
- Firearms offences.

The UK team set to the task of identifying, tracing recovering and protecting 274 victims of child trafficking by this gang. The aim which is ongoing was in a multi-agency operation with police specialists, local authority children's safeguarding experts and NGOs. At the time of the report, only 81 had been recovered and supported. Sadly, many have been moved out of the country by the OCN and likely replaced with fresh victims that are unknown to the authorities.

9 Strategic Achievements

- Primary in the set-up of the Home Office inter agency working party on trafficking of children
- Advising the 'London Child Safety Board' and writing contributions to their 'Toolkit for identifying trafficked children'
- Achieving the first UK conviction of Trafficking Human Beings (THB) of a child
- Achieving the second conviction of an offence in the UK of 'internal trafficking'

---

[11]They were still awaiting trial in 2016.

- Providing written evidence to Lady Butler-Sloss (member of the 'All Parliamentary Committee on THB') that has been passed to the Attorney General (unedited) raising the issue of lack of sentencing guidelines to Judges in THB cases not linked to sexual exploitation
- Advising the United Kingdom Human Trafficking Centre (UKHTC) and the Serious Organised Crime Agency (SOCA) (Knowledge) on Roma organised crime
- Advising NPIA[12] and contributing to the revised Association of Chief Police Officers (ACPO) 'Child Abuse Manual' and revised 'Guidance for International Investigations'
- Operation Golf currently has a seat on the ACPO Child Trafficking Working Party
- Superintendent Gravett and Chief Inspector Carswell acknowledged by SOCA, Europol and EU as the only UK Police JIT experts and have been involved in training SOCA, Europol and UNODC staff.

Operation Golf concluded on the 31 December 2010; however, officers of the Metropolitan Police Financial Investigation Unit continue to identify and seize criminal assets from the Tandarei gangs. In 2013, they sent 24 International Letters of Request to the Romanian authorities to trace and seize criminal assets from the Tandarei gang.

The OCG were disrupted to a significant degree. However, the problem has not gone away. There are always more gangs who are willing to exploit the vulnerable of both adults and children from Europe's poorest communities. The Tandarei gangs were active in every European country and beyond. They targeted crowded places, tourists, the vulnerable and elderly and women.

In 2011, having been appraised of the case at the European Parliament, the European Commission issued a new directive 36/2011, which acknowledged that the use of victims for the commission of low-level crimes such as begging, theft, pick pocketing and other crimes is a form of labour exploitation. This opened up the definition under the Palermo Protocol of human trafficking and brought it in line with modern forms of exploitation and into the modern context.

Superintendent Bernie Gravett retired in April 2011 after 31 years as a dedicated police officer. He continues to work for international organisations to train law enforcement, border guards and NGOs in combatting human trafficking. This is a crime that generates millions and affects thousands of lives.

Police need to be proactive, seeking out the suspects and organisations and networks. They cannot rely on victims coming forward and presenting themselves and their stories to the police.

---

[12]The National Policing Improvement Agency (NPIA) was a non-departmental public body in the UK, established to support police by providing expertise in such areas as information technology, information sharing and recruitment. NPIA existed from 1 April 2007 until 7 October 2013 (Source: https://en.wikipedia.org/wiki/National_Policing_Improvement_Agency, accessed 16 October 2016).

We know where the crimes emanate from, and where and how the victims are exploited. We just need the will to go and find them. Human trafficking and is a criminal business operation and needs to be seen as such. By acknowledging that children have a value in their income generating ability, police and society must look at child begging and child criminality differently. In some EU countries, begging is not a crime, this is an outdated philosophy. In modern countries with state support and the availability of education, no child should be on the streets begging. They should be in education being given opportunities to improve themselves. Such opportunities did not exist in the past but they do now. Law enforcement, government bodies and other actors in the piece need to investigate every act of begging and criminality by children and look at who is 'pulling the strings'.

Open source research is an invaluable tool in identifying the presence of OCN activity across Europe. As a postscript, Superintendent Gravett and Chief Inspector Carswell went on to for another JIT with the Bulgarian authorities. This focussed on the trafficking of girls and young women for the purpose of pick pocketing. They identified over 3000 girls and women moving around the world committing such crimes. All of whom passed through London at some point. In one case, open source investigations revealed that a girl, who had been arrested and convicted of pick pocketing crime in London, had also been arrested and convicted in Australia, Austria, Switzerland and Germany. The information was gleaned from newspaper crime section reports revealed in open source analysis.

Radicalization in a Regional Context: Future Perspective on Possible Terrorist Threats and Radicalization

Holger Nitsch

1 Introduction

In the aftermath of the attacks in Brussels, Paris and Copenhagen, the growing threat of an unspecific kind of terrorism is within Europe and its people. Fear has risen, and the result has been that more radical parties did get votes in the last election periods within Europe. The offenders were radicalized young people which didn't grow up in countries like Afghanistan, Libya or Lebanon. These individuals grew up in the local neighborhoods in Europe, which doesn't necessarily lead to terrorism, but there might be some driving factors that could indicate that these events are more likely to happen in the future. That's why it is important to research on their motives to counter the threat and estimate the likeliness of these events to happen again. Most of the papers in Europe are talking about the action to be taken and that is in most cases a repressive action. Unfortunately, there are not many voices calling for research of the background, the reasons and the grievances. On the other side, researchers around the globe are trying to find solutions to minimize the threat and to detain people to become radicalized and maybe even terrorists.

Of course, terrorism is not an entirely new phenomenon. It exists since centuries, most likely since mankind developed different ideas of societies. What has changed is the methodology and the modus operandi. The main objective of terrorism is still to put fear in the hearts and minds of citizens, and in order to reach that goal it can be observed that most attacks are not taken place in western states, but that the attacks in the West have a much higher attention in the media and therefore also this number is rising. As in the nineteenth century and most of the terrorist actions in the

H. Nitsch (✉)
Department for Policing, College for Public Administration in Bavaria, Bavaria, Germany
e-mail: Holger.Nitsch@pol.hfoed.bayern.de

twentieth century targeted institutions or representatives of the states or the economy, now the main target is to hurt and kill as many citizens as possible by attacking them during their normal way of life. Airplane hijackings first wanted to get pressure on states to negotiate about prisoners or other political demands. In the next step, airplanes were blown with bombs and since 9/11 it is a fact that they can be used as a weapon. To protect a society from attacks like in Paris or Brussels seems to be impossible without giving up fundamental rights of freedom, self-determination and democracy.

What is it that drives people to become terrorist, to choose a way of life outside the society, to choose another society and to disrespect the right to live for many other individuals? That is best described as the process of radicalization. Unfortunately, there is no common background for being vulnerable to radicalization. The European Expert Group on Violent Radicalization states that "one of the most significant understandings gained from academic research over the recent years is that individuals involved in terrorist activities exhibit a diversity of social backgrounds, undergo rather different processes of violent radicalization and are influenced by various combinations of motivations. This is relevant not only with respect to the more recent expression of Islamist terrorism but also as regards right-wing, left-wing and ethno-nationalist manifestations of such violence previously experienced in a number of European countries" (European Expert Group on Violent Radicalisation 2008). The origins of terrorism do not always have their grounds in political grievances. The causes might be very personal. There are some facts which can be generalized, despite the fact that there are and always will be exceptions.

Most later terrorists started their journey toward extremism between the age of 14 and 25. People at that age are looking for their independence and their own identity outside the parental home. If they are dissatisfied with a situation, they are looking for a solution and it is very easy to be unsatisfied. The world itself is not just, and never has been, and in every society an individual finds inequity and social injustice. If the feeling is strong enough to believe to change that situation and the feeling of powerlessness is big, the individual might find some easy sounding solution from an extremist side. Usually it starts with a discussion about any topic of grievance and the offer to join the radical, but nonviolent group. This would be the first step into radicalization. Nowadays, this can be done much easier with the possibilities of communication through the internet and Web 2.0 applications.

Many young people have the feeling to do something against injustice and basically to make the world a better place in their eyes. The drivers to radicalization and the motivations into extremism might be very different and are very individual. In fact many studies prove that there is no common reason or common background for people to become radicalized (Sageman 2004). In their teens, many young individuals have the idea of changing the world, but the most of them will never get the idea of using violence or getting on the path of radicalization.

2 Understanding the Drivers of Radicalization

To understand the drivers of radicalization, it is necessary to have a short look at the ideologies behind it. Extremist ideas do attract people mainly by giving simple solutions to complex situations. This could be global warming and who is guilty for that, animal rights or whatever. Extremist ideologies can be divided into several categories, which are dominant for the main narratives. In reality, there is very often a mixture between different ideologies. In the following, there will be a rough explanation of possible extremist or radical ideologies that could lead also to violent acts and terrorism (Nitsch 2001, p. 300ff).

Ethno-, national Extremism: Groups in this category believe to fight for an ethnic group or for independence of a region within a nation. Examples from the past are The Irish Republican Army (IRA) or the Basque Euskadi ta askatasuna (ETA).

Revolutionary Extremism: Very often called also left-wing extremism, but the phrase revolutionary is more precise. The aim of groups following this ideological path is to get the population to a revolution. They don't see themselves as head of the new socialist or communist system, but as the spearhead of the coming revolution. In Europe, violent active groups have been the German Rote Armee Fraktion (RAF), the Italian Brigate Rosse (BR) or the French Action directe.

Right-wing Extremism: Groups adopting this ideology have a wide range of ideological elements. Most of them see their nation as something better than others and xenophobia is a key element. American white supremacists believe that the white race is superior, which is also adopted by some European groups. The depreciation of people of other origin makes violent acts easier, and these are grounded on the ideological background. A Europe-wide acting group is Blood and Honour, but there have also been terrorist groups like the German Nationalsozialistischer Untergund (NSU), who killed people just because they were foreigners and also one police officer as a representative of the state.

Religious Extremism: Here is a great variety of different ideologies. It can be summarized that there is no religion that cannot be abused and misinterpreted as a justification for violent acts and extremist views all over the world despite the cultural background. In Africa, there is the Lord's Resistance Army (LRA), which sees itself Christian, or Boko Haram as an Islamic terrorist group.

Sects: They have also a religious background, but not all sects can be assigned to a religion. Violent acting sects do have in common that their followers believe that the end of the world is near and the last battle—Armageddon—between the good and the evil is soon to come. Their members see themselves as the chosen ones to survive. One example here to be mentioned could be Aum Shin Rykio, the Japanese cult, that committed the Sarin gas attacks on the Tokyo subway in the mid-nineties.

Environmental Extremism: Protecting the environment as an issue or animal rights groups are of course not extremist organizations, but there are groups that are willing to use violence in order to get closer to their aim. The issue that can be a single issue or a complex of issues regarding ecological topics might even justify injuries or the death of innocent people. A very active example of a terrorist acting group is the Animal Liberation Front (ALF).

Idiosyncratic/Lone Wolf Extremism: Usually idiosyncratic ideologies are believed by single individuals, because these ideologies are created by the experiences of an individual and what the influences are. These ideologies are neither homogenous nor do they have to be comprehensible or logical. They might contain many elements from other ideologies, but then this will be changed and adopted to create their own "truth." Examples would be the UNA bomber in the USA, the Bavarian Peoples Front in Austria (one individual that hated Roma and Sinti) or Anders Breivik in Norway.

State Terrorism: A state can be involved in two ways in extremism. If the ideology of the state is from a democratic perspective an extremist ideology (like the systems of North Korea or in Cambodia of the Khmer Rouge), then the state is using violence to stay in power and to oppress its citizens, like it was happening in Cambodia, and also in many other states in history. The other possibility is that the state gets directly involved in violent acts by either supporting them or directly planning and committing violent acts in hostile countries. Examples would be the bombing of the PA 121 in 1988 also known as Lockerbie or the bombing of the discotheque LaBelle in Berlin. Both perpetrated by agents of the Ghaddafi regime.

Within the above given framework, most of the radical ideologies can be summarized. Many contain a revolutionary element that fits perfectly to the world of ideas of young people and the process of uncoupling from their parental relation. Giving the feeling of being revolutionary is a good and powerful driver toward extremism. It can be observed that left-wing organizations use the term "revolutionary" as well as right-wing and also Islamic groups. The right-wing scene has adopted this element as it proved to be successful for the recruitment of young people in the left-wing scene. So phrases like "world-revolution" can be seen on posters on demonstrations, and the phrase is used in skinhead and other right-wing orientated music, which is very important for the recruitment of new members of the scene.

The ways to radicalize individuals are very different and differ also by the ideology behind the process. There might be the possibility that research might be able to filter factors or indicators that lead to radicalization and hereby help to develop future preventive measures as it was described already by various authors. For example, Nesser (2004) identifies within the violent radical group four types of personality: the leader, the protégé, the misfit and the drifter. This classification helps to identify different ways of radicalization for these different types of personalities. This is most probably comparable with the way to become a member in an Organized Crime Group. The methodology of research might stay the same, just some indicators and driving factors should be changed.

3 Radicalization as a Phenomenon with a Regional Context

Looking at radicalization as a phenomenon with a regional context, it is obvious that there are regions, where the radicalization process is easier than in other areas. Taken the area north of Baghdad in today Iraq, it is quite easy to find out why there is quite a huge sympathy for the IS in a city like Fallujah and why, even with similar background data, this is not the case in other regions. While the feeling of not being represented by the government in Baghdad is quite strong within the Sunni population as in Basra, a Shiite city in the south of Iraq, this feeling is not represented in the government, being dominated by Shiites. On top of that, the regime of Saddam Hussein was dominated by some Sunni clans. Many leaders of the so-called Islamic State (IS) are former members of the Iraqi military under the rule of Saddam Hussein. This easily explains the support of the Islamic state by the Sunni population in Iraq, but not necessarily the support in Syria and absolutely not the wave of foreign fighters coming from Europe and other countries and regions to Syria to fight for IS.

Also scientific research is needed regarding the growing number of lone wolf terrorist actors in Europe and the Western World. The lone wolf phenomenon is also strongly connected to the strategy of leaderless resistance that was promoted as a concept by the Klansman Lois Beam in 1992 (Burton 2007). This strategy aims to form small groups, with no or very low hierarchy, to act flexible against the state. The ideology shall be the driver, and it should not be under the command of a greater movement which would be interstratified with informants. These small groups can act independently and will be very flexible and hard to detect. This can be boosted by breaking the action down to a single person who will be even harder to detect. This is also a reason for the difficulty to fight lone wolf terrorists, connected loosely or not at all to a radical group, but sharing the same or at least very similar ideology.

Dalgaard-Nielsen (2010) has put the most relevant facts of what we know about radicalization together, but she shows also which information we are lacking off and which information is needed, where there is a need for further research.

As an example, the focus could be on two (four) cities, but for sure it would be working also in many other European cities or regions.

Belgium has the highest number of foreign fighters per capita in Europe, and many of them are coming from the town of Verviers, which is a middle-sized town. The question is: Why from Verviers and not from another city of the same size and similar sociological, economical, demographic and political conditions? Verviers has close to 60,000 inhabitants. The economic situation is not very promising: shops are closing and the unemployment rate is quite high. The number of migrants is also quite high (Urbistat 2011). In other cities, like Mons or Liege, there are more inhabitants, but the problems are the same. They are in the same region, the unemployment rate is similar, shops are closing, there are many migrants, but just a few or none foreign fighters from these cities (Eurostat 2016). It can be estimated that radical preachers that visited Verviers have also been in Mons or Liege.

Another example is the city of Dinslaken (Urbistat 2012). It has around 70,000 inhabitants and is home to the most foreign fighters in Germany. Many believe that this is connected to the fact that the number of migrant in Dinslaken is very high—especially in the quarter of Lohberg. While Dinslaken is economically not prospering, it is not suffering. On the other side, there is the city of Offenbach with 120,000 inhabitants and home of the most inhabitants with a migration background in Germany, compared to the total number of inhabitants (Stadtverwaltung Offenbach 2016). The interesting fact is that there are also just a few or none foreign fighters from Offenbach and also here, like in Verviers, Mons or Liege, there are similar economical or sociological backgrounds.

The question resulting from these two short examples is: What are the drivers and factors there that are getting individuals on the path of radicalization? What is known is that grievances, the economic situation, a strong feeling for injustice and cultural uncertainty are strong indicators for radicalization, but they do not explain the difference in the number of radicalized individuals in the various prescribed regions.

4 Foreign Fighters and Homegrown Terrorism

The phenomenon of Foreign Fighters seems to attract people from all over the world to help IS fight in Syria, Iraq and Libya. The estimated number of foreign fighters early May 2016 is between 3922 and 4294 individuals from the European Union. Most of the fighters come from the United Kingdom, France, Germany and Belgium (with the highest percentage per capita). A great uncertainty is about the percentage of converts which ranges from 6 to 23%. Over 90% are coming from an urban or a built up area. 17% are female and 14% are confirmed dead. 30% have returned home (Van Ginkel and Entenmann 2016, p. 4). Will these 30% be a threat to their country of origin? Obviously the answer "yes" or "no" would be far too simple. There are returnees that pose a threat to democratic societies and some who quit with the ideology and even might be traumatized by what they experienced in Syria, and some might just disengage. The threat is posed by returned fighters that are still convinced to be able to change the world and that their path of violence is the only way to change it. Also these might be heroes in their local, regional scene. So they might be a role model for others in their region to become more radical. Cecilia Malmström, the European Commissioner of the DG Home Affairs has pointed out clearly the threat emerging from foreign fighters:

> Often set on the path to radicalization in Europe by Extremist propaganda or by recruiters, Europeans travel abroad to train and to fight in combat zones, becoming yet more radicalised in the process. Armed with newly acquired combat skills, many of these European 'foreign fighters' could pose a threat to our security on their return from a conflict zone.

And as the number of European foreign fighters rises, so does the threat to our security (Malmström 2014, p. 4f.).

Another phrase that made it into the media is "Homegrown" extremism or terrorism. This expression includes the belief that extremism is very often coming from the outside. The phrase is mostly used in the context of Islamic terrorism. For a long period of time, people in Western societies intended to believe that Islamic terrorism is happening in Iraq, Lebanon, Israel or in states of central Asia. If the West was a target, then it was mainly either in one of the above-mentioned countries or regions or through a group that infiltrated into the West and committed an attack there. The Hezbollah mainly acted in Israel, but the hijacking of the Achille Lauro was committed outside the usual acting area. 9/11 was perpetrated by people from different countries, but none was a citizen of a Western state, but the four pilots were radicalized in a mosque in Hamburg. The attack on the Olympic Games in Munich 1972 by the Black September was coming from "outside." Therefore, in the minds of the citizens, many violent acts with an extremist background are coming from outside the country and maybe the offender have even another cultural background and are intruding the usual way of life. An exception has been the acts of the Irish Republican Army in the UK and of the ETA in Spain, but for the most countries this was the case. These were homegrown cases. Also acts of violent left- or right-wing groups are homegrown, but they weren't named like that. The phrase describes a phenomenon that the radicalization of individuals with that ideological background is not expected to happen here, because the ideology seems not to fit in the cultural area. That is why the term is used on Islamic extremism in Europe. People don't expect individuals to adopt a radical Islamic ideology within the cultural context of the West. This leads also to the fact that citizens are more frightened of Islamic extremists than of other, besides the fact that the media coverage plays an important role here as well, which will be covered later in this article. But it is important to understand that not just Islamic radicals can be homegrown, also terrorists like Breivik, Fuchs or the members of the Baader, Meinhoff group are homegrown. That could help to find a way to a systematic and analytic research of the homegrown radicalization phenomenon.

5 Drivers of Radicalization and Means of Recruiting

In the past, strong drivers of radicalization have been the general political situation or socioeconomic reasons. For left-wing groups in the 1970s of the twentieth century, the so-called 68 revolution and the Vietnam War were such drivers. Further, the personal perspectives and the unemployment rate drove many young people into right-wing radicalization as it was the case after the breakdown of the German Democratic Republic. The lack of alternatives of youth culture is another strong driver. If there is just one ideological possibility for young people in a certain area, individuals intend to join the group to belong somewhere. To belong to a peer group is of utmost importance for young people.[1] The tools to recruit new members were personal contacts, leaflets, music and papers.

The invention of the Web 2.0 has changed the possibilities of recruiting drastically. Now the range of radical offers has multiplied. There are many new ways to get information to the individual, and it is harder to judge for the individual which information is correct and which one is incorrect. All levels of radicalization can be covered easily by Web 2.0 applications, and it is much harder to counter the radical narrative. It can be observed that the radicalization process is not starting with the most radicalized ideas, but with certain personal needs. Right-wing groups, as well as Salafi-Jihadi groups, try to place friendship and comradeship in the center of their ideology and that attracts young people looking for exactly these very much. The possibilities of influencing are visualized in Fig. 1.

Young people do not seek for radical ideas or ideologies. They are looking primary for the feeling of togetherness (Roy 2007, pp. 53–60; Malik 2015). Hence, the outsider, the one who does not belong to a peer group, is more vulnerable to radical ideologies—not because of the radicalism of the ideology, but of the feeling of belonging to a group. This vulnerability is a key element for young people to get on the path of radicalization. Besides, they have the feeling that they have to change something in the society. They believe that they are able to live a life of self-determination within the rule of the radical ideology, and they will not see the contradictoriness of this believe. They might be driven into radicalization by personal feeling of deprivation, conflicts or other problems. In their adolescence, individuals do not belong to their parental environment the same way they did, when they were children. On the other hand, they do not necessarily belong already to a certain peer group; they try to find their own ways and solutions for the problems they are facing or they think are important. During this phase of life, it is easier for radical groups to recruit individuals, because of the crisis of identity of young people. That gap could be filled with the feeling of togetherness topped with radical ideologies.

6 The Radicalization Process

Usually, the individual follows during the process of radicalization also a change in his own use of language. It radicalizes as well. The radical narrative, like any narrative, is often a matter of interpretation. Based on the communication theory of Schulz von Thun, every information that goes from the sender to the recipient has four aspects (Schulz von Thun 2016):

---

[1]Author's interview with Manuel Bauer, who was a right-wing activist in the 1990s, who stated that there were just right-wing oriented in his area and that there was no alternative. He would have joined other groups, if there would have been the possibility.

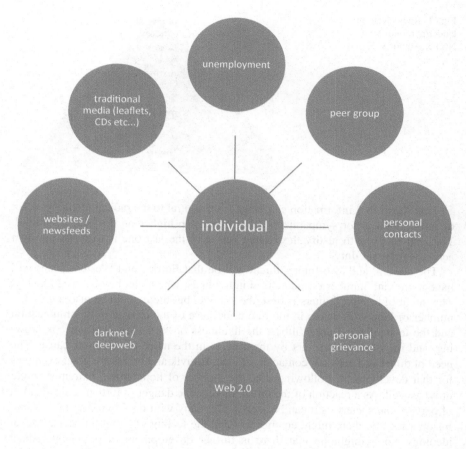

Fig. 1 Possibilities of influencing and radicalizing individuals (author's display)

Relation: It tries to build up a relation with the recipient to persuade him and to touch him personal.

Self-revelation: This means that the information given is authentic, and the sender gives something from his own views to share it with the recipient.

Facts: The facts contained in the information might be in this context ripped out of context to prove the point.

Plea: The information also requests from the recipient some kind of action. This could mean just to reconsider mainstream opinion, further action or change the own way of thinking.

The narratives are for the different stages also on different levels. The drivers might change during this process. The deeper the individual might be dragged into the ideology, the more communication might be taken place via deep web or dark web. In the Dutch National Coordinator for Counterterrorism (NCTb) report, Schmid (2010, p. 60) visualized it as shown in Fig. 2.

Fig. 2 Radicalization ladder/pyramid. *Source* NCTb. Author's display

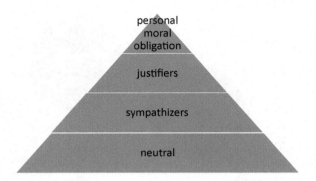

This shows that information can range from neutral to the radical ideology. A lot of information is for sympathizers who might step higher on the ladder/pyramid. The next block is to justify the violent acts, and the last one is to make the own action a personal duty.

The current, and also future, threat will be that Europe and Western states will have a growing number of radicalized individuals, which can be observed already. The threat of foreign fighters is described above, but there could be also a growing number of lone wolf actors in the future, because of the change of communication and the technological possibilities, the likeliness of "self-radicalization"[2] is growing, and more people will act by themselves in the name of whatever cause. The need of direct and personal contact has faded. Breivik and others are good examples for that development. Following this, the number of homegrown extremist might grow as well. As a reaction of the threat, there is the danger of further radicalization of parts of the society as it can be seen in Germany with the PEGIDA[3] movement. Inside a society, there might emerge fear and the feeling of being threatened by an ideology, so it might be that there is further development of opponent radical ideologies. As the extremist organization knows these mechanisms in a society as well, it can be interpreted that this mechanism of counter-reacting can be used for radicalization purposes as well. Two of the attackers of the second Paris attack were registered twice as a refugee—one time in Greece and the other time in Serbia. In the refugee situation with this amount of people coming through the Balkan route and many of them not being registered, it seems to be very unlikely that an individual gets registered, if he wants to avoid that (Howell 2015). The purpose is that in the aftermath of the attack all sources will be contacted and that there is a strong feeling of refusal against refugees. As IS knows that mechanism as well, it might be done on purpose to rise the tensions between the societies in Europe and refugees.

---

[2]This term is not precise, but used in the media. The individual is acting alone, but influenced by material and narratives read and consumed on the internet or Web 2.0 applications.

[3]PEGIDA is the acronym for: „Patriotische Europäer gegen die Islamisierung des Abendlandes"/ „patriotic Europeans against the Isamization of the Occident".

7 Possible Future Radicalization Scenarios

The methodology of case studies is very useful to research on the phenomenon of radicalization, because not all the information will be there, like in real life. Every case will have certain ways of looking at it, and there is no easy or single cause solution, but it helps to get a full picture for the decision making process on how to tackle the problem (HBS 2014). Following scenarios will illustrate possible future scenarios of radicalization and risks to democratic and free societies within different ideologies. These scenarios are based on research on existing cases which are adopted to future possibilities. For the reason of anonymization, they are abstract and no names are given:

Scenario 1

An individual with a migration background from Turkey is well integrated in the society. He has a couple of school mates he spends his time with. His family is supporting him with his school career and is supporting him for his studies after school. He is brought up and living in a middle-sized city with several schools and also a small university in a rural area. One of his best mates with no migration background is getting radicalized with the idea of Salafi Islam and dreams if going to Syria to fight for IS and die as a martyr. As they spend much time together, he adopts the ideas of this ideology as well. They joined the same martial arts club to train their fitness and become ready to fight abroad. They share videos and sites on the internet. His friend decides to go to Syria and dies there. As he is following his friend's steps, he is also reading posts how his friend is describing his life in Syria and is fascinated by the way of living. When his friend died in combat, the messages about his death stated him being a hero and fighting until the last round to really deserve to go to paradise. In the aftermath, he hopes to follow his friend to go to Syria and to join the Islamic State to die as a martyr like his friend did. He posts his ideas in several social media sites and also states that he would always follow the rules of his understanding of the Islamic law[4] and also act against his family, if they are violating his understanding of that code.

Relevant elements for his radicalization process in this scenario are the peer group and the strong relationship with his best friend. At a later stage, Web 2.0 applications play also an important role as this is a way to keep communication up and running and get the "saga" of heroism in the battlefield. Furthermore, via this channel the daily life can be reported that there is a meaning of life and that the individual is recognized as an important person there.

---

[4]Islamic law: sharia. Very often in extremist Islamic groups, the interpretation of that very complex issue is very much simplified.

Scenario 2

Anna[5] has just started her studies in a new city. She has regular contact with her family which lives in another region. She takes her studies seriously and makes some good friends with her fellow students. Anna is vegetarian and so are many of her friends. As there are reports about the torture of animals in several institutes, she discusses this within her peer group and some of her friends decide to join a demonstration and so she takes part as well. She has not been active before, but she likes the people she meets there and that there is the feeling of a common cause that could change things to the better in her eyes.

As a next step, she decides to engage more and she starts collecting reports on animal cruelty and industries that use animals for testing series. Whenever there is something to do, she more and more voluntarily offers her help and engagement in distributing leaflets or taking part in demonstrations or similar actions. Her contact with her family still is present, but she starts to speak about the subject of animal cruelty more and more. Her fellow students, and her family start to ask for other topics to discuss and other things relevant in life, but she keeps up the same topic over and over, so that her friends from school and her fellow students start to avoid contact with her. She is not attending all courses at the university she should, and tutors and professors asking for her get reluctant answers. She found her cause and her whole life is starting to concentrate on that single issue. Even the contact with the group organizing demonstrations and other actions gets lower, as she starts to believe that they do not do enough to solve the problem. She is getting her information on the Internet and collects all information she can on that issue.

Anna finds other people on the internet who are thinking the same way she is, and she makes contact with them to join this group. At this time, she strongly believes that violent action is the only way to stop animal cruelty. She stops to communicate with her former friends, and the communication with her family gets very low.

Key elements of Anna's path to radicalization are a mixture between a peer group, which was important at the beginning and later the information she believes to be true on the internet. Communication with Web 2.0 applications will get her deeper into extremism and her willingness to act violent will rise, despite the fact that the people who originally got her on the path had a clear idea where to stop the action and what is inappropriate for the cause. This scenario is also a classical example for the mixture of the influence of social media and peer groups. What is very obvious in this case is that there is a point during the radicalization process where the communication with former friends and maybe even with the own family stops or at least changes significantly. This can be called a point of no return, because the individual tries to break down former bonds to live a new life with a new family.

---

[5]Based on a scenario created within the EU-funded project ISDEP: www.isdep.eu.

Scenario 3

Karim is well integrated in the society. He has a job at a local transport company doing shift work and meets his colleagues from work from time to time. He is not married, and his family is not living with him, but in his country of origin. In his time off, he regularly surfs through the internet, and from a certain time he comes over videos that supposedly show allied forces commit crimes to people in Iraq. He starts to search for more material and he finds it. In the real world, his contacts with his colleagues get less, and he more and more digs himself into the quest to find more material to prove that the West is committing crimes against people in the Middle East. Following this, he starts building up his own view of the world based on the material he is searching for. He makes some contact with a fake account with other like-minded persons around the globe who think that the West, and especially the USA, is guilty for the injustice and the current unfair situation of the world. He never meets these people personally, they are just sharing the same thoughts and discuss open what to do against this injustice. Karim starts to think about getting active by himself and informs himself about easy ways to get access to weapons. Then, he actually is starting to plan something, but very unsure what it will be. In Karim's case, there is a trigger that he watches a video on a YouTube channel that shows a war crime. He decides that he will give his answer in the form of a violent act the next day.

This scenario shows a lone wolf scenario, where an individual radicalizes himself. This is not quite correct, because there is the support of the internet and social media. Nobody had the possibility to discuss his grievances with him, because all his communication was anonymous in the web. This scenario can be changed easily to the case of a more sophisticated offender like Anders Breivik or other terrorists acting alone.

The key driver in this scenario is social media, which led to self-radicalization without physical influence of a third party. This influence is just virtual, but the result is a person, who created his own view of the world and got himself trapped in a very personal and unique view on how to resolve what he sees as a problem. It proves as well that there is a lack of possibilities to counter radical narratives in this case and that at this stage there are just very few ways to detect the individual earlier to prevent a violent action.

8 The Media, in Particular the Social Media, as Key Facilitating Factors

The three scenarios are just examples on how different radicalization might take place and that the ideologies are exchangeable. What is obvious, and can be observed in many researches, is that *the role of social media is growing*. It cannot be stated that it is the key element or how big the influence is, but there is an effect and that is growing and measurable (Bundeskriminalamt 2015).

Following this, there is another key element of radicalization missing and this is the role of the media. It would be quite easy to blame media for the problem of radicalization, but if there is a common understanding of how media works, what its duties and objectives are, and how it can be manipulated, it can be used for radicalization as well as for de-radicalization for countering violent extremism and for disengagement strategies. It is also the platform to offer alternatives where radicalized individuals state that there was a point where they were lacking that alternative to prevent them to go the next step ahead.

Besides the classic media, like newspapers, leaflets, magazines, radio, TV and so on, the internet and social media changed the world of gathering and sharing information rapidly. The amount of people worldwide with access to the internet, and by that also to social media, was growing enormously during the last years. Most of the refugees entering Europe in 2015 did not have much, but they had at least a smart phone, which was and is absolutely necessary for them. On the one hand, they can stay in contact with friends and family left back home, and on the other hand they could gather information on the best ways on how to enter European soil, despite the fact that this information, often provided by crime groups smuggling people, was often wrong.

8.1 Radicalization and the Role of Social Media

All types of media give information, but, unlike the classical media enterprises, the internet gives everybody the possibility to share their views and perspectives with everybody, who wants to, through blogs, websites or social media. It might be compared with an anarchistic surrounding, because rules and regulations do exist, but are very often not followed, and matters of enforcement are weak or missing because of the way the internet is organized. With the classical media channels, it was easy to judge the information in order to what the consumer knew about the media enterprise. This is no longer so easy possible with social media. It is much harder to validate information as correct or true than before, and this is a problem that makes it easier for radical ideologies to get more recruits. There is an offer for every level, every stage of the radicalization process and very often in many different languages. As it was very unlikely for young people in the West to join an Arabic terror cell, that picture has completely changed with al-Qaeda and IS, and the use of the internet for propaganda purposes in various languages and mainly in English.[6] Extremist organizations are fully aware of the importance of the media for their propaganda.

To prove that point there will be a couple of examples: Breivik did not want to die and he was very keen not to be called insane in order to get his court case to use this as a stage for his "show." Luckily, the Norwegian society and the court were

---

[6]Both organizations have their own e-zines: Inspire for AQ and Dabiq for IS.

mature enough that it did not work out what he imagined in his delusion to happen: that there will be a revolution against the existing system.

Most of the people born in the 1980s or before know what they did on 9/11 and how they first heard about the attacks on the World Trade Centre and the Pentagon. The media wing of al-Qaeda even produced a video of the attacks filming the crash into the WTC from a spot in Brooklyn to publish it through their media channels, and they did produce videos to commemorate this attack on anniversaries (Site Intel Group 2010). Even nowadays, if there is an attack, where the connection to Islamic terrorism is not proven, there is a high likeliness to broadcast pictures of falling towers.

With the intense growth of social media, the effect of propaganda multiplies by posts on Facebook or Twitter. After every attack, there are people writing posts, why they think this is good and others, who think that there is an easy solution for the problem by eliminating the other side. This can be observed by many attacks, and some EU-funded projects did some research on that.[7] As internal communication of radical groups might make more and more use of the possibilities of the darknet or the deep web, the offensive and radical narratives have to stay on the surface to spread the ideology for propaganda purposes, and the trust of many users in the anonymity of the internet makes the effect of hate and violating messages rise. These feelings have to be fed over and over again; otherwise, they will be overcome by other ideologies. This propaganda is working the same way as advertisements are. The input has to come regularly than it is easier to keep the followers on track. Once persuaded, the input has to come to keep the individual at a certain level (Cialdini 2006).[8]

8.2 Interdependencies of Media and Terrorist Incidents

The interdependency is illustrated in Fig. 3.

As terrorism could be the final step of the radicalization process, this illustrates the interdependencies between the different actors involved. The aim of the incident is, on the one hand, to put fear in the heart of the citizens of the affected society, and, on the other hand, to gain further support from support groups or even state sponsors. Even if the citizens are not direct victims of an attack, it will be very likely that, if the incident took place on a public space, a subway or similar, like in Paris, Copenhagen or Brussels, that citizens are afraid of going out in public and that the feeling of insecurity has risen. To increase this effect media coverage plays an important role. All types of media will report about the incident, and everybody

---

[7]SMART CV is one example for a project that specifically dealt with the countering of radical narratives after terrorist incidents. Another project where the issue was dealt with was ISDEP: www.isdep.eu.

[8]Cialdini describes different aspects of persuasion. He refers not to radical propaganda, but the tools and the psychology is transferable.

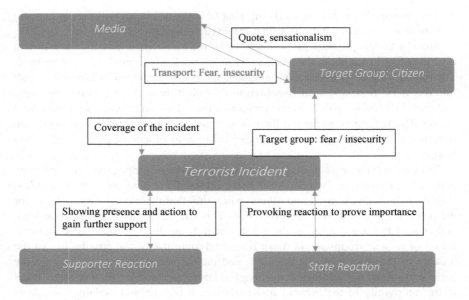

Fig. 3 Interdependencies between the different actors involved in a terrorist incident (author's illustration)

will try to get the "best pictures," which are usually the most horrifying ones. Alternatively, it has to be a symbol like the London bus. During that attack, the bus was in most of the media coverage, because it gives a better view of what has happened than pictures from the also attacked subway, despite the fact that more people died due to the subway bombs (Independent 2015; Spencer 2012; Jenkins 1981).

Moreover, these emotions are multiplied by the simple fact that we, as humans, want to see it. We do not turn away from our media source. If we have to go, we follow the situation on Twitter, because we want to hear about the latest developments. As we are afraid and shaken in our security, we expect the state to act in some way, and that forces state authorities to create new laws or even to invade in another country, like in Afghanistan after 9/11 and later in Iraq. Therefore, the system of interdependencies is quite simple, but unfortunately working for the extremists.

That leads to the conclusion that effective attacks have to be against Western targets because the media coverage and the resulting affects are stronger. Citizens do not remember the last attack in Baghdad, because that moved from front page news to the end of the paper, because we got used to it, and it is far away in the imagination of most of the population. That this might be connected to threats at home is not obvious for the most and not present, which is also not to be seen as negative; otherwise, a normal life in a free and democratic society would be hard to live.

9 Conclusion

Considering all of the above, radicalization will most probably not stop or lower, but rise. It can be expected that not only big events, like football matches or festivals, are targets. To hit citizens' emotions in their daily life, the threat will target also daily routine like it happened in Paris and Brussels. The kind of perpetrators might be similar: a small group or a lone wolf actor, acting on behalf of whatever ideology. There is the danger that the franchise system of al-Qaeda will not only be copied by IS, but also by other groups with a different ideology.

To tackle the phenomenon of radicalization and violent extremism, there is the need of a multilevel and a multi- and interagency approach.

Further research is needed on the causes of radicalization, the drivers and indicators that lead individuals on the path of radicalization. That includes the evaluation of the vita of radicalized persons to find indicators and drivers, an exchange of information of data concerning this issue between research institutes, state authorities and different countries. A deep analysis of the PESTLE factors[9] seems also necessary, considering also sociocultural aspects. As the ideology seems not to be as strong with some extremist group members as it used to be in the past, also here is research necessary to get a clear view of the attractiveness of that radical movement.

The interagency approach should include all key stakeholders to try to prevent as many individuals as possible to join the path of radicalization. This ranges from schoolteachers, university lecturers, social workers, prison staff and many others, to one of the key players—law enforcement agencies. They will be the ones, who have to deal with an incident, if it is too late, and if a problem is detected or assumed, people intend to tell it first to the police.

For sure, there is not one solution to the problem. As the problem of radicalization is very complex, there has to be also a variety of solutions, meaning that there should be a variety of offers to people to get alternative views not to become radicalized, or different possibilities to leave a radical group and to de-radicalize, and for some there has to be a variety of possibilities to disengage.

One possible approach to find out more about drivers and indicators is the EU-funded project ePOOLICE that developed a prototype that enables law enforcement agencies to find new indicators and trends in organized crime and terrorism (ePOOLICE website 2015). This will work on radicalization as well, if the idea and knowledge is picked up and brought further. This methodology, which is used also in other projects, is similar to the grounded theory. The result is known: a violent extremist. Now it is important to find, based on the knowledge available, new indicators and drivers to better understand the process of radicalization.

Lowering the risks of radicalization and violent extremism should be the main objective for all parts of the society and state institutions and could be done by a multilevel and multiagency approach to make free and democratic societies safer.

---

[9]PESTLE stands for: political, economical, social, technological, law and environmental.

References

Bundeskriminalamt. (Ed.). (2015). *Analyse der Radikalisierungshintergründe und -verläufe der Personen, die aus islamistischer Motivation aus Deutschland in Richtung Syrien oder Irak ausgereist sind*. Wiesbaden.
Burton, F. (2007). *The challenge of the Lone Wolf, Stratfor*. https://www.stratfor.com/challenge_lone_wolf. Accessed April 12, 2016.
Cialdini, R. (2006). *The psychology of persuasion*. New York: Harper.
Dalgaard-Nielsen, A. (2010). Violent radicalization in Europe: What we know and what we don't know. In *Studies in conflict and terrorism* (Vol. 33). London: Routledge.
ePOOLICE Website. (2015). www.epoolice.eu. Accessed May 08, 2016.
European Expert Group on Violent Radicalisation. (2008). *Radicalisation process leading to acts of terrorism*. http://www.clingendael.nl/sites/default/files/20080500_cscp_report_vries.pdf
Eurostat Website. (2016). www.europa.eu/eurostat/de. Accessed April 22, 2016.
Harvard Business School HBS. (2014). *The HBS Case Method*. http://www.hbs.edu/mba/academic-experience/Pages/the-hbs-case-method.aspx. Accessed May 26, 2016.
Howell, K. (2015). *Two men linked to Paris attacks registered as refugees in Greece; 129 dead, 352 wounded*. Washington Post online. http://www.washingtontimes.com/news/2015/nov/14/two-men-linked-to-paris-attacks-registered-as-refu/. Accessed May 5, 2016.
Independent. (2015). *7/7 bombings: Who were the 52 victims of the London terror attacks?* http://www.independent.co.uk/news/uk/home-news/77-bombings-london-anniversary-live-the-52-victims-of-the-london-terror-attacks-remembered-10369569.html. Accessed June 5, 2016.
International Centre for Counter-Terrorism—The Hague. (Ed.). (2016). *The Foreign fighters phenomenon in the European Union*. The Hague. http://icct.nl/wp-content/uploads/2016/03/ICCT-Report_Foreign-Fighters-Phenomenon-in-the-EU_1-April-2016_including-AnnexesLinks.pdf. Accessed April 24, 2016.
Jenkins, B. (1981). *The psychological implications of media-covered terrorism*. Santa Monica: Rand Corporation.
Malik, K. (2015). *Radicalization is not so simple*. https://kenanmalik.wordpress.com/2015/10/07/radiclization-is-not-so-simple/. Accessed May 31, 2016.
Malmström, C. (2014). Cited in: AGENFOR Italia (2014). *Foreign fighters*. Published within the framework of the EU funded project EURAD.
Nesser, P. (2004). *Jihad in Europe. Exploring the motivations for Salafi-Jihadi terrorism in Europe post-millennium*. Oslo: Department of Political Science, University of Oslo.
Nitsch, H. (2001). *Terrorismus und Internationale Politik am Ende des 20. Jahrhunderts*. München: Inauguraldissertation, Ludwig-Maximilian-Universität.
Roy, O. (2007). Islamic terrorist radicalization in Europe. In S. Amghhar, A. Boubekeur, & M. Emerson (Eds.), *European Islam—Challenges for society and public policy* (pp. 52–60). Brussels: Centre for European Policy Studies.
Sageman, M. (2004). *Understanding terror networks*. Philadelphia: University of Philadelphia Press.
Schmid, A. (2010). The importance of countering al-Qaeda's 'single narrative'. In The National Coordinator for Counterterrorism (Ed.), *Countering violent extremist narratives*. The Hague.
Schulz von Thun: Das Kommunikationsquadrat. http://www.schulz-von-thun.de/index.php?article_id=71. Accessed April 29, 2016.
Siteintelgroup. (2010). *Al-Qaeda Video commemorates 7th Anniversary of 9/11*. http://news.siteintelgroup.com/blog/index.php/about-us/21-jihad/55-commemorate. Accessed May 2, 2016.
Spencer, A. (2012). *Lesson learnt—Terrorism and the media*. Swindon: Arts and Humanities Research Council.
Stadtverwaltung Offenbach. (2016). *Einwohner mit Migrationshintergrund*. https://www.offenbach.de/medien/bindata/of/Statistik_und_wahlen_/dir-18/dir-29/BEV3-2015_-_Migrationshintergrund_Migra_Pro_seit_2009_Grafiken.pdf. Accessed April 25, 2016.

Urbistat. (2011). *Maps, analysis and statistics about the resident population, municipality of Verviers.* http://www.urbistat.it/AdminStat/en/be/demografia/popolazione/verviers/63079/4. Accessed April 22, 2016.

Urbistat. (2012). *Maps, analysis and statistics about the resident population, municipality of Dinslaken.* http://www.urbistat.it/AdminStat/en/de/demografia/dati-sintesi/dinslaken/5170008/4. Accessed April 22, 2016.

Van Ginkel, B., & Entenmann, E. (Eds.). (2016). *The foreign fighters phenomenon in the European Union. Profiles, threats & policies* (Vol. 7, No. 2). The International Centre for Counter-Terrorism—The Hague. http://icct.nl/publication/report-the-foreign-fighters-phenomenon-in-the-eu-profiles-threats-policies/. Accessed April 26, 2016.

Why Do Links Between Terrorism and Crime Increase?

Luis de la Corte Ibáñez and Hristina Hristova Gergova

1 Introduction

In accordance with the Royal Decree 873/2014 approved by the spanish Council of Ministers last 1 October 2014, the Ministry of the Interior will create a sub-department of the State Secretary for Security following the integration of the structure of the Spanish National Centre of Antiterrorist Coordination (CNCA) and the Spanish Intelligence Centre against Organised Crime (CICO). The duties assigned to the new Spanish Intelligence Centre against Terrorism and Organised Crime (CITCO) are to boost and coordinate integration and, the assessment of fees, reports and operational analysis disposed by law enforcement bodies concerning terrorism, organised crime and violent extremism develop criminal intelligence strategies in this regard, establish action criteria and operational coordination between concurrent organisations and, design global strategies to combat the phenomena previously named.

Alongside concern to optimise resources and remove administrative duplicity, the Spanish Ministry of the Interior provides as a further justification, in order to create the CITCO, the existence of direct linkages and objectives between terrorist activities and individuals involved in other criminal activities as warning about growing similarities in several action patterns carried out by terrorist groups and other criminal organisations. Taking into account the implications of the previous and the relevance of the institutional and strategic change, which will contribute to the creation of the new body, it seems like the right moment to examine the problem of interaction between terrorism (or associated insurgents phenomena) and other

L. de la Corte Ibáñez (✉)
Universidad Autónoma de Madrid, Madrid, Spain
e-mail: luis.cortes@uam.es

H. Hristova Gergova
Private Sector, Madrid, Spain
e-mail: hristina.gergova@gmail.com

criminal activities, especially organised crime. This article tackles the nexus terror-crime and the various insights in the field, its possible forms and, the main criminal activities involved. Hereafter, special attention is given to the causes that enable and prompt terrorist involvement in criminal acts and relationships, outlining a model about them and offering an additional revision of the endogenous and exogenous factors which can boost such links. After that, we review the consequences that could arise from the convergence between terrorism and common and organised criminality. Finally, some conclusions on this matter are drawn.

2 A Possible and Increasing Nexus

For decades, the hypothesis on the existence of substantive linkages between terrorism and other types of criminal acts, in particular common and organised criminality, has been observed with scepticism (Bovenkerk and Chakra 2004). The most obvious reflection of that attitude refers to the traditional organic separation of departments and units with power in antiterrorism and criminal investigation, as well as the theoretical and academic approach equally differentiated from each of the problematic areas referred to. In apparent congruency with such approaches, but still admitting the possible connection between both and including some occasional indications and proofs, the EUROPOL annual reports on terrorism and organised crime have not been able to identify any final trend on this matter.

However, for several years now, that conceptual and organisational separation has started to be questioned by an increasing number of researchers and experts. For the time being, in a Resolution (1373) adopted a few days after the 9/11 attack in 2001, which took place in New York, Washington and Pennsylvania, the United Nations Security Council (2001) pointed out:

> (The Security Council) notes with concern the close connection between international terrorism and transnational organized crime, illicit drugs, money laundering, illegal arms-trafficking, and illegal movement of nuclear, chemical, biological and other potentially deadly materials, and in this regard emphasizes the need to enhance coordination of efforts on national, subregional, regional and international levels in order to strengthen a global response to this serious challenge and threat to international security.

The same sort of warnings will reappear in other further strategic documents issued by the United Nations, other international organisations and numerous countries. Accordingly, the National Security Strategy approved by the Spanish Government in 2013 suggests that the confirmation of growing links between criminal and terrorist organisations entails a substantive increase in the level of hazard of both phenomena for Spain's security. Such statement matches several conclusions revealed one year before in a specific report commissioned by the European Parliament. The authors note to have fund wide evidence on the existence of a sort of "marriage of convenience" between terrorist groups and organised crime inside the European Union countries.

Among the major arguments and evidence that nowadays support a reassessment of the relations between terrorism and other criminal forms the following stand out.

1. The mentioned relation is not natural or necessary, nor is it impossible or unnatural. It is proved in several past and present examples from terrorist actors (networks, big organisations, small groups or individuals) involved in the performance of other crimes. In fact, according to the data provided by the Ministry of the Interior, between 2005 and 2011, 20% of imprisoned individuals in Spain for alleged association to Jihadist elements previously went to prison as a consequence of their involvement in other criminal activities (Díez 2015). There are also examples, although less numerous, of actors related with organised crime that have perpetrated actual terrorist attacks and violent crackdowns.
2. While the connexions between terrorist actors and petty crime has been relatively frequent in different sceneries, the convergence between terrorism and organised criminality uses to be much more frequent and widespread in countries and region affected by weak institutions, as the so-called fragile or failed states.
3. In accordance with an increasing number of experts, the lack of evidence that could endorse significant links between terrorists and organised criminality in developed and stable countries may not be the consequence of the inexistence of this type of links, but to the compartmentalised treatment that its security and intelligent agencies grant to such threats. After all, it is considerably more difficult to find evidence about the convergence between two criminal phenomena when, for instance, the agencies in charge of investigating them do not consider plausible the confluence between the two, nor have they been incentivised to cooperate and share information about these issues.
4. The analysis of some of the most important terrorist attacks committed in the last fifteen years in occidental countries, considered as genuine strategic surprises, has revealed contacts and connexions with common and organised crime that if they would have been taken into consideration, they might have eased its prevention. As an example, we can mention the two most serious terrorist attacks committed in occidental countries in recent decades. Two of the 9/11 hijackers were aided by criminal facilitators to obtain the visas they needed to stay in the USA. Other two hijackers obtained the driver's licences that they used to board the aircraft through members of a criminal network operating in northern Virginia (Shelley 2015). Also, several perpetrators of the Madrid bombings on 11 March 2004 had a criminal career previous to its radicalised phase and used their criminal contacts to finance the operation and obtain the explosives required to perform it (Reinares 2014).

3 Types of Convergence and Illicit Activities Involved

The convergence between terrorism and crime may introduce changes in the nature of the entities and groups that characterise it, as well as in its activities (Makarenko, 2004). Furthermore, those changes can adopt diverse forms and degrees.

The first way of interaction between the phenomena discussed involves the direct and independent implication of terrorist actors (whose violence normally responds to political or political-religious objectives) in criminal practices whose commission allows economic inputs and other material resources to be obtained. Or, vice versa, the commission of typically terrorist attacks (that is to say, aimed at frightening and coercing wide audiences) by groups or criminal structures with a lack of ideology. As we can see, in both cases the convergence consists on the *appropriation* on the part of an actor of the modus operandi characteristic of the other.

The *cooperation* between terrorists and criminals is an other option, not necessarily incompatible with the first one. Such cooperation can respond to two different motivations. The predominant ones are of a practical nature, and they lead to the development of transactions or agreements for the exchange of goods and/or services. This category includes buying and selling weapons, explosives, documentation or any other resource the terrorists may be interested in and that can be offered by criminal elements. A part from supplying illicit goods, the clients or partners from the criminal world can provide to the terrorists the access to shelters or routes that allow clandestine transit from some countries to others or support services related with transfers and laundering of illegal funds or bribing public officials.

Vice versa, terrorist can also sell services to the criminals. In some weapon, drug or even human trafficking operations the terrorists play the role of sellers, either because they are becoming mere criminal actors or because they use that sort of goods to exchange them for some sort of useful resource of the terrorist army, like when drugs are exchanged for weapons. Likewise, terrorists can rent their intimidation and violence potential for other criminal purposes, for example to protect drug crops and illicit goods. Not all transactions between terrorists and criminals are chosen by the latter. Indeed, sometimes those transactions are promoted through some kind of extortion pattern. This is the case of some of the taxes that the extremist groups impose upon illegal traffickers as a requirement to cross a territory under their control. For example, recently Islamic State has used its control of a swath of land on the central Libyan coast to profit from the smuggling of migrants (The Soufan Group 2016). The cooperation can also involve the joint participation in one or several phases of the illicit trafficking chain or the subcontracting or commissioned services (from document falsification to laundering proceeds). Furthermore, it can cover from the realisation of opportunistic transactions posed on the initiative of either actors to building tactic partnerships that imply the joint collaboration in violent actions or campaigns, either because of ideological affinity or to anticipate a common benefit obtained from such alliance. Occasionally, such alliances can result in a long-term relationship of a parasitic or symbiotic nature.

The previous two types of convergence can take place in an occasional, sporadic, continued or even permanent basis. However, the more successful experiences of lasting involvement or cooperation can evolve towards a partial or total *transformation* of terrorist priorities. At some point, a terrorist group can become a hybrid structure between a terrorist group and a criminal organisation, so motivated by money than by ideology and politics. For some observers, this could be illustrated by the progressive involvement in drug trafficking of the insurgency of FARC in Colombia (Chernick 2007). It can even totally depoliticise the real objectives of their activity, despite to keep an ideological facade. Some analysis has suggested that this could be the case of the Uzbekistan Islamic Movement (UIM), a jihadist organisation that in the middle of the 2000's controlled 70% of the routes used to export the Afghan opium and heroin to Central Asia (Cornell 2006).

And, at least theoretically, a criminal organisation could also incorporate political motivations to its activity or end up becoming an activist actor. Thus, the so-called D-Company, an Indian criminal organisation running by Dawood Ibrahim, a top criminal boss of Bombay for many years, has been involved in several campaigns of attacks on Indian soil since the 1990s, all of them developed in cooperation with various jihadist groups, including the Pakistani organisation Lashkar-e-Tayyiba.

There are numerous illegal activities through which terrorism and crime can converge. These include, among others: illegal trafficking of different nature (drugs, weapons, humans, natural and energy resources, precious stones and various goods), burglaries and sale of stolen goods, kidnappings, extortion, illegal services of protection of people and products, fraud and scam and money laundering (see Table 1).

Terrorist actors can become involved in those activities through different modalities of criminal convergence previously mentioned. The ones connected with illicit transnational markets, such as the drug, weapons and human ones, inevitable result in a certain amount of cooperation with purely criminal actors. Others, like the burglary operations aimed at acquiring a supply of weapons or other logistic useful resources, certain extortion practices, the kidnappings or money laundering, can be developed independently, but this does not exclude the option of cooperating with individuals and criminal specialised structures which can even be subcontracted. Finally, the implication (independent or not) in those activities with a capacity to contribute with maximum revenues (manly the drug trafficking, and other illegal trafficking and the systematic kidnapping practice) is the most likely to

Table 1 Main criminal activities related to Organised Crime/authors' illustration

Illegal trafficking	Drugs, weapons, humans and natural resources (precious stones, oil, wood, other); Illicit smuggling of several products and goods (tobacco, fuel, food and goods considered basic necessity, counterfeit and stolen products or exported/imported illegally)
Other	Burglaries and sale of stolen goods, kidnappings, extortion and blackmail; illegal services of protection of people and products, fraud and scam; sexual and work exploitation; maritime piracy; cybercrime, money laundering

open a process of criminalisation/de-politicisation of terrorist actors which, as we have already seen, may be partial (hybridisation) or total (transformation).

4 The Terrorist Involvement in Activities and Criminal Relations: Towards an Explanatory Model

Since the participation of terrorist actors in crime activities and/or criminal relations is not a constant, it is simply an option, it is important to explain this possibility as the opposite (that is to say, it is also important to explain the cases where the criminal implication does not take place or is missing). Like any other modality of action, the probability that a conventional terrorist actor becomes involved in criminal actions or alliances, leaves them or refrains them will be determined by *ability*, *opportunity* and *motivation* factors. One could say that the first two (ability and opportunity) are essentially objectives, while the third one (motivation) is rather subjective. To offend or to get in contact with criminal elements, you need to have certain capabilities and opportunities that make it possible as without them criminal involvement would simply be impossible. However, the previous is not enough, because in order to become involved, the criminal activity must be chosen, normally according to its utility; in short, it must be desirable and desired by its own protagonists. Let us analyse the issue in depth.

Let start with the matter of the *capabilities*. Some of the basic skills associated with the terrorist activity are usable for the commission of some kind of crimes. The habit of proceeding clandestinely, the violence potential and the lack of scruples when the time to use force comes, it provides the terrorist with a clear advantage, when the time to undertake certain criminal practices comes like extortions, burglaries, kidnappings and prevent that these activities come out into the open. Instead, the possibility of becoming involved in more complex and sophisticated criminal activities or unrelated with the use of force is subordinated to the availability of capabilities or abilities specific for its development. Some of them would be more available for terrorists groups that have among their members individuals that come from the crime world and are still in touch with it. Such contacts, for example, are normally essential to enter the markets related to illegal trafficking. Furthermore, the big terrorist structures, endowed with an heterogeneous militancy, powerful capture and recruitment instruments and wide support social bases also have more options for the incorporation of individuals with the necessary experience and training to perform specialised criminal activities, like document falsification, carrying out embezzlements through internet or financial transactions of money laundering. Nonetheless, the absence of the required capacities for the independent development of certain illegal activities may be compensated through the cooperation with some specialised criminal agents: counterfeiters, drug growers and producers, traffickers and smugglers, swindlers, thieves, hackers, etc.

The realisation of criminal activities is also a matter of opportunities. After all, no all the scenarios and circumstances are equally enabling for the criminal action and the establishment of collaborations or criminal alliances. This situational conditioning largely explains why the geographical distribution of convergence cases between terror and crime is not equal and reaches its highest frequency and intensity in countries and regions less stable, as mentioned in a previous section. The opportunities of being criminally involved can significantly vary according to other elements, such as the efficiency of safety systems established in the scenarios, where terrorists operate, the access to natural resources or goods that could be commercialised in some illegal markets or the presence/proximity of criminal actors in those same scenarios.

But, naturally, the criminal involvement will only take place, when a direct motivation towards it exists. We can assume that in either cases, it is a matter of instrumental motivation, dependent upon the utility the terrorist actors themselves attribute to their involvement in criminal activities or relations. In particular, the main utility of extracting from the commission of criminal actions, characteristic of minor and organised crime, is obtaining economic and/or material resources useful for satisfying the demands raised by terrorist activities. Between such demands appear the ones related to the preparation and execution of attacks, the acquisition of all means and equipment required to do this (weapons and explosives, vehicles, documents, costs for transfers and travelling, communication technological gadgets, computer scientists, etc.). Due to the low cost of terrorist conventional operations, some economists have correctly defined terrorism as a sort of "cheap war" (Valiño et al. 2010). However, terrorist groups still have to confront a broad range of expenses related to the creation and maintenance of their operating and support structures. There is also the money destined to the upkeep of the militants (accommodation, maintenance and other basic expenses), the tasks of canvassing and recruitment of new members, the training (sometimes in camps that need to be build, run or rent or move) and, propaganda, among others. As a consequence, some terrorist organisations, the most powerful and ambitious, develop budgets entailing annual expenditure amounting to hundreds of millions of Euros or dollars. The amount of economical resources that a terrorist organisation disposes of determines its capacity to act and, therefore, its hazard. Furthermore, in comparison with other forms of financing terrorism, obtaining funds through illegal activities presents some advantages. First, through their involvement in illegal activities terrorist can access quickly to the needed revenues. Second, some illegal actions can be used at the same time to get money and also to provoke certain social and political effects (social control over the extorted population, the overtaking manoeuvre and duress towards a State in case of kidnapping). Third, the terrorist groups whose funds come from a reference community or a sponsor State are more dependent that those who financed themselves by illegal means. Finally, the involvement in certain illegal business particularly profitable can allow accumulating benefits that largely exceed the money that is necessary to satisfy the operational and organisational requirements.

However, criminal involvement also entails risks and costs. The illegal actions by the terrorist actors and their contact with other criminal actors may facilitate their entrance to the security and intelligence agencies' radar, risking the detection of their militants, operations and structure. Likewise, in the event of its criminal involvement transcending, the terrorists would be exposed and give an image of marginal or "mafia" organisation that could make them lose social support among its reference community. The criminals who get involved in deals with terrorist groups or join their ranks could become informers for the security and intelligence agencies, either because they fall under the control of the agents or because they offer try to sell them compromised information. Finally, the economical benefits of collaborating with professional criminals are enormous, but they also may contribute to corrupt militants and reduce their ideological commitment. For their part, their criminal partners the moment they agree on cooperating with terrorists, not only do they risk the discovery of their illegal business, but also the control over them (the extremists with whom they collaborate could end up owning them), a part from confronting a possible increase of the judiciary and police pressure upon them and more severe penalties to their activities. Lastly, we can add to the above, strictly speaking, several costs which result from the involvement in crime. Like, for example, time, energy and resources (human, material) that such involvement may consume and not been used in other tasks directly related with the terrorist activity. The expectations of such risks and costs explains, why many terrorist actors, that do not find an alternative to criminal involvement to finance their activity, chose to commit a minor crime and void collaboration and competition with other criminal structures.

In short, the estimates that any terrorist actor may carry out in order to determine the convenience of getting involved (or continue its participation) in activities or criminal relations will involve a combined evaluation of benefits, risks and costs. Leaving aside the cases of full depolitisation/criminalisation, terrorist actors are usually less willing to risk their security in exchange for greater economic benefits than other criminal actors. For this reason, and given the importance of risks and costs detailed above, presumably a terrorist actor will only be sufficiently motivated to become involved in criminal activities or relations under de following two conditions:

1. *The need*: when the criminal involvement allows resolving an economical or logistic necessity that cannot be satisfied otherwise.
2. *The sufficient advantage*: although the basic economic necessities are covered by nonillegal means, terrorists can involve in some lucrative illegal activities at low risks conditions without exposing them to any kind of cost or risk that they would not be able to assume.

The distinction between a motivation based on the necessity and one inspired by the advantages related to the criminal involvement is relevant, because they have different intensities and vary in relation with the capacity, opportunity and risk factors, examined above. Undoubtedly, the appearance of a strict necessity to become involved in criminal activities to enable the terrorist action constitutes a more powerful impulse than the mere search for advantages. Thus, when a terrorist actor has no other choice, but to become involved in criminal actions or deals, the capacities and opportunities that may lack to do this will be created and looked for, besides having a much higher willingness towards risk. Instead, when the criminal implication becomes only one option among others, their choice would be probably much more conditioned by the available capacities, the opportunities that happen unexpectedly (not necessarily looked for) and, also the magnitude of the predictable risks and costs.

Summing-up, the participation of a terrorist actor in criminal actions and relations requires a decision, whose development will be conditioned by the capacity, opportunity and, motivation factors that we have just examined, in accordance with a deliberative process whose sequence could match the one illustrated in the following image.

Once identified in the abstract the general factors that determine the possibility of terrorist actors becoming involved in crime should be noted that the expression of each of those same factors has no fixed, unique and equivalent value for all cases, but instead those can vary significantly depending on the moment and the characteristics of each actor. That is why we can detect differences relevant in the terrorist proclivity towards crime.

5 Enhancers of the Current Convergence Between Terrorism and Crime

Although the involvement of terrorist elements in criminal activities is not new, growing evidence suggests a progression in this regard. Therefore, we still need to identify the factors that could promote the trend, what we could call the *drivers* of the convergence between terrorism and crime.

In theory, any change in the probabilities of criminal involvement by one terrorist actor should reflect a variation in one or several of the explanatory dimensions included in the model sketched in the previous heading. Thus, the increase in frequency and degree of terrorist involvement in criminal actions requires that the involvement itself becomes (Fig. 1):

1. *More necessary*, to the extent that previous alternatives for covering the economical, material and any other type of requirement disappear.
2. *Easier*, if the capabilities and/or required opportunities to offend or make contact with criminals.
3. *With more advantages*, either because criminal opportunities arise that allow obtaining inaccessible resources in any other way (for example, certain kind of armament) or increase its income in a substantive manner, what may constitute a difficult temptation to resist.
4. *Less expensive and risky*, as often happens when operating in certain scenarios or environments characterised by a high level of criminal impunity.

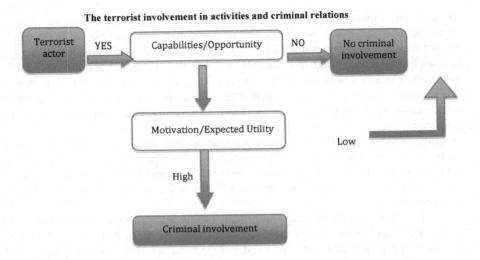

Fig. 1 Terrorist involvement in activities and criminal relations

There are several factors that can induce the previous changes. Some of them can respond to the own development of the terrorist and criminal activities involved, being thus of an *endogenous* nature. On the contrary, others will be *exogenous*, related to elements and external forces prone to influence in the means and conditions in which those phenomena express and operate. This is described below.

5.1 Endogenous Factors: The Fostering Connection Between Evolving Terrorism and Crime

To some extent, the confluence between terrorism and crime (mainly on their organised crime patterns) is due to the variations in the own nature of those phenomenon. The following are some of those endogenous factors.

The international spread of terrorism and organised crime. Among all the tendencies that have marked organised crime evolution, the international spread has probably played a decisive role (UNODC 2010a, b). Whereas all organised crime may not concern cross-border scope, for the most part it does and continues spreading, because of the following main causes: the substantial increase of collaboration between groups and criminal organisations from different locations, the emergence of illicit global markets, whose negotiating phases operate throughout the world and the outbreak of criminal organisations with an active presence or erected at an international level. Terrorism, yet it has not reached organised crime benchmarks, it has met a growing and progressive internationalisation for its part, but on an uneven manner. For that matter groups and organisations operating within

Al-Qaida and the global jihadist movement network stand out among the rest of existing forms of violence performing as international terrorism main exponents. Organised crime and terrorism internationalisation is a step towards the cross-border expansion of their activities, structures and objectives across countries, regions and continents. This process' consequences increase the range of meeting points in incentives and opportunities between both phenomena. Perhaps the most important would be to create various illicit markets at international level. While its economic implications deserve an apart remark, it becomes noticeable that a global market's configuration related to illegal arms trade greatly facilitates the access to such fundamental resources for terrorists and ensures a very specific interest to contact and carry out deals with criminal components. Another significant effect is the fact that terrorist actors coincide in a growing number of countries and scenarios, which is a potential condition for possible collaborations. Furthermore, the terrorist's rising need to move across multiple borders may turn them into preferential customers for criminal networks involved in counterfeiting identity cards, as well as in facilitating the illegal transit of migrants in various parts of the world.

The expansion of global criminal economy. The responsibility for the massive increase of profits resulting from illicit business noted during the last decades (Naím 2006) is held on the progressive interconnection between actors and groups related to organised crime and the resulting emergence of criminal networks at regional and global level. According to UN estimates based on 2009 data, organised crime might be generating gains up to 870 of billions of dollars per year, amounting to 1.5% of global GDP, to 7% of a year world export of goods incomes and to the annual GDP of a country such as the Netherlands, whose economy appears among the top twenty of the planet (UNODC). Such an increase in their profit can continue and indeed increases its attraction, offering terrorist actors its possible introduction into global criminal economy circuits, which explains for example the progressive connection in some scenarios between terrorism and drug trafficking as the first source of incomes related to cross-border organised crime, as well calculated by the UN around 320 annual billions of dollars. Another consequence of the emergent global criminal economy that must be undermined is its close relation with the proliferation of tax havens and fiscal bodies and financial engineering technics applied for illicit money laundering, which criminal advantage can be shared with other actors, including terrorist organisations.

Frames decentralization and multiplication process. During last two decades criminal and terrorist groups have met a similar evolution regarding their organisational features. They have diversified their models and increased the differences between them in terms of size, or number of members and structures. There has been a transition from the big organisations prevalence with a hierarchical structure and a highly centralised decision-making system (quasi military in some cases) to the combination of the latter with other organisational models and the proliferation of smaller groups, with a more flexible and dynamical functioning and more decentralised structures, including minor networks practically horizontal, some of which have been created spontaneously to undertake just one or few activities but likely to collaborate with other structures from bigger bodies or to be integrated in those. There are various reasons that can explain why the participation of terrorist in criminal activities can turn out to be easier, therefore more likely to happen, for terrorist groups whose structure and features resembles more to the new organisational models more horizontals and smaller in size. First place, the greater horizontality or status equivalence between terrorist' groups members the greater the freedom of choice each one of them will have and the less control the leaders will exert. In such circumstances, the probability for prevent individual involvement in criminal activities becomes low. Furthermore, the group's capabilities to raise money are usually proportionate to its size, especially if these funds are seek by nonillegal means, such as individual o collective donations, legal business activities or state sponsorship. For this reason, sometimes smaller terrorist groups could be less resistant to the temptation to get involved with illicit business.

The onset of autonomous cells and individual terrorism. During the 80s, some ideologists from the white supremacist movement in the USA of America proposed among their supporters a new way to perform their violent activities with racist background that would break with the traditional terrorist groups and organisations characteristic functioning. Those leaderless strategy defenders proposed to take advantage of the new opportunities offered by information and communication technologies to foster their violent supporters to act on their own initiative, at the same time avoiding the maintenance of organisational structures that would help safety agencies and judicial authorities to make connection between them. More recently, some big jihadist organisations (including Al-Qaida) and some of their most important strategists decided to contribute and reproduce a guideline, as the one that has been described above, to exploit its great communication and propaganda potential to foster their supporters to perpetrate terrorist attacks on their own, wherever possible (Sageman 2007). The strategy in question explains the present autonomous cells and individual actors proliferation, which are responsible for a growing number of terrorist attempts in Europe and North America, as well as these actors frequent engagement with criminal practices, it is not a surprising tendency, for which the reduced size of these actors lacking of any kind of external financing source.

Their relation with constituencies and their level of popular support. The changeability of positions taken by different terrorist groups on the criminal matter, or by the same group in different periods of time may be strongly determined by its constituency or community of reference and their characteristics. Groups with large popular support may be financed by their financial donations and contributions, which removes the criminal attempt that many of these groups conceive as the main incentive. Besides, this dependency may also act as an obstacle for the terrorist engagement in criminal activities as far as it may evoke dismissal or disappointment among the supporters that compose the respective collective. But this may result in very different ways if business and illicit activities to develop must also be of benefit to the terrorists respective collective, as it happens with drug trafficking business in countries or regions, where great part of the population have no alternative to barely cover their living expenses or to obtain a decent salary. Speaking more generally, terrorist groups lacking of the support of big collectives whose contributions are insufficient will have more freedom to choose the methods needed to cover their economic and logistic demands or to take advantage of illicit business. Finally, an unexpected dramatic decrease of the funds provided by their supporters may boost criminal activities.

Intersections between terrorism, organised crime and armed conflicts. Terrorism is an activity that can be applied to many different scenarios and can as well find intersections with war, where it may undertake an auxiliary role as a complement to other forms of violence, as guerrilla or conventional warfare. These combinations have gained relevance during the last years due to interactions between jihadists groups. Some of the latter with such capacities and size that have become authentic militias or small and private military forces and local insurgencies (Islamists, nationalists or seculars). Since the end of the last century, war between states, due to boarder armed conflicts, has been substituted as the main war trend by organised crime, which now has the main role among many war economies and a significant influence in the evolution of conflicts and in postconflict environments (De la Corte 2013a, b). The coexistence of uprising actors/terrorists and criminals within war environments creates junctures especially conducive to links between both phenomenon. This is an issue that appears themed as "hybrid wars", a concept frequently used in strategic and military analysis (Hoffman 2007). Criminal and terrorist element inevitable contact and their connection to armed conflicts' dynamic stimulate its connection through other ways. In the first place, insurgencies practicing terrorism are opened to a bigger range of criminal activity opportunities. Just as it happens, when they achieve to take control of a territory or they enter trafficking circuits that determine to large extent war economies. In the second place, criminal organisations can draw great profit from their deals with terrorists/uprising actors, obtaining their protection as well as becoming favoured customers. In the third place, in such scenarios both actors would take advantage of the impunity that emanates from the fragility of the State, to pursue efficiently their criminal activities. Above all this, criminal organisations may be interested in sustaining the conflict's continuity supporting or collaborating with terrorists/uprising actors.

Other upheavals, terrorists' groups and their members, have to deal with. Terrorist's participation within criminal activities might prosper, because of elements related to the inner life and the terrorist actor's trajectory. As mentioned before, the existence among the ranks in a terrorist group of one or more militants with criminal backgrounds enables the group's implication on illicit activities. In recent years, an increase in radicalisation cases performed by individuals coming from the criminal world, with particular incidence in some European countries, has been noticed (Nesser 2015). Turning to a different subject, sometimes the only reason to take a terrorist group or organisation to get involved in criminal actions is their leaders or command structure's opposition to the action itself. This is why the second environment, that may foster criminal implications, is the emergence of leadership crisis. This would be the outcome of the detention or elimination of some of their leaders, or the consequence of the emergence of an inner conflict between confronted leaders, groups facing opposite attitudes related to the convenience of involvement in criminal activities. Finally, the last situation that may encourage the terrorist involvement in criminal activities is the entrance into a phase of decline or demoralisation. Normally, as terrorist's organisation members loose fate in the cause, they orientate their lives towards new objectives and there is an increase in dropouts, which at the same time tends to aggravate organisations crisis. Deserters, lacking of other skills than the one acquired during their militant phase and with limited opportunities to return to an ordinary and legal life might feel tempted to use their experience as terrorists in a new criminal career. Terrorists avoiding to abandon their activities, even when they have assumed the impossibility to achieve the political objectives that once motivated them, might find the same way. Therefore, during the last phase of certain terrorist's group active life (that in some cases may last for years) or even after its official dissolution, it is not unusual that one of its sectors survives as a simple criminal group or as mercenary forces (Cronin 2009).

5.2 Exogenous Factors and Favourable Environments

Terrorism as well as crime (especially organised crime) has demonstrated great ability to adjust to changes within their activity areas and new business opportunities resulting from the contemporary societies' evolution and the succession of political and economic circumstances and connections The influence that many among those elements has in terrorism and crime connections has already been explained in the preceding section, which details and identifies endogenous factors, that in the end cause other external or exogenous phenomenon.

State's sponsoring decrease. During the last decades of the Cold War terrorism and other forms of subversive direction violence functioned to some extent as a foreign policy instrument to some countries. This was possible because of the economic support and sponsoring these countries' States provided to big range of terrorists actors: terrorists groups and armed militias from the extreme left, nationalists, Islamists, groups from the extreme right, etc. Nonetheless, especially the fall of the Berlin Wall and later the attacks of 11 September 2001 ended the generalised resort of State's sponsoring, partly because of the arising geopolitical changes and also because of the growing dismissal public opinions against terrorism. There is a broad consensus considering this trend as the main inductor to the increase of terrorist's participation in criminal activities, as an alternative to the financial and logistical support previously received by one or other State (Hutchinson and O'Malley 2007).

Response to terrorism and their means of financing. As it is well known, another important consequence of the attacks of 9/11 was the launch of global counter-terrorism fight, which was first headed by North American authorities, but immediately followed by specific response actions and programmes planned by multiple international organisations and an long list of countries. UN Security Council Resolution 1373 (2001), mentioned before, called on all Member Countries to implement and enforce new measures to prevent and restrain financing terrorism. Above all, it was requested to type crime as: providing or collecting funds deliberately to prepare terrorist actions; freezing immediately funds and other financial actives or economic resorts of everyone related to terrorist activities; the denial of harbouring to those who were financing them. Urging as well to implement general international regulations stated under the forty Recommendations on money laundering and the nine special Recommendations on terrorism financing of the FATF. Based on these directives, which included an explicit condemnation to State's sponsoring practices, the fight against terrorism financing became one of the fundamental dimensions in the counter-terrorism field. However, we can assure that even if these actions were necessary and efficient, they have had different results and consequences on the relation between terrorism and crime. On the one part, the direction to detect suspicious connections between legal economy flows and terrorist groups and organisations have reduced the opportunities to obtain funds from States and powerful private contributors, whose stocking and reception might require the intervention of financial intermediate entities unprincipled. On the other part, when disabling those financing channels another powerful incentive is added to seek financing by means of criminal operations participation and illicit business in all their forms (Shelley 2015).

Destabilising trends, armed conflicts and transition process. There is no need to explain the consideration of terrorism as a destabilising factor to countries and societies that suffer it, in deed, the effect defines the objective pursued by the proper terrorists. Organised crime ability to induce instability, despite not being their main objective, also becomes obvious when criminal groups get involve with intense campaigns of violence (sometimes purely terrorists), aimed against their competitors or authorities and the Sate (UNODC 2010a, b). It has already been remarked, the growing tendency of nonstate armed actors using terrorism in war environments and managing their economy getting involved in multiple illicit markets and interacting with criminal organisations and networks. However, it is important to repeat that to notice the way the essentially destabilising nature of terrorism and other expressions of organised crime do necessarily bring previous and external destabilising trends. On the other hand, the specific advantages armed conflicts provide to terrorism and crime interaction are to a large extent similar to those in initial phases of certain political transition process, weather they have been preceded by violence or not. Lastly, all the situations referred above can discharge chaos and might erode the ordinary functioning of States' services, or even in extreme cases void them, which certainly brings great advantages to their opponents.

Access to (micro) environments enabling criminal activity and collaboration. Not all contexts are equally favourably to the junction between terrorism and crime.

Nor all provide the same opportunities and incentives, nor all are equivalents regarding their costs and costs. Certain scenarios are especially propitious for criminal activities growth, due to their physical, sociopolitical and human characteristics. Therefore, once the direct collaboration with actors from the criminal world (individuals or groups), the connection between terrorism and crime will be deeply related to contextual nature factors. After all, collaboration requires a common point between those who will cooperate, to recognise common objectives and to establish a sort of relationship of trust between them. Some of the environments where these contacts are more likely to arise are:

- *Prisons*, mostly lacking of a separated restriction system for terrorists, and therefore permitting extremist recruiters to attract and radicalise prisoners.
- *Marginal neighbourhoods* such as the ones existing in a large number of big and capital cities, lacking of institutional presence, affected by high levels of crime and deeply entrenched on a criminal subculture, to which it is frequently added a great concentration of immigrant population or representatives of minorities.
- *Cross-border areas* in which there is a conflict regarding national jurisdictions due to the intersection of laws and therefore function as a transition point to international illicit trafficking and terrorists.

Critical scenarios. Probabilities for convergence between terrorism and crime also tends to vary depending on some countries and regions and others, with great advantage for those whose terrorist's actor's presence would be in line with one or more than one of the following risk circumstances (including some already explained before) (De la Corte 2013a, b):

- The availability of crop's areas related to the manufacture of drugs or natural sources susceptible to illicit exploitation.
- Root connection between one or more trafficking joint points between manufacturer's countries and consumer's countries.
- Armed conflictive risks.
- Political transition process.
- State's fragility or collapse (State's and social services absence).
- Economic and social underdevelopment.
- High level of corruption and black economy.
- Large and porous borders.

Finally, if various of the latter coincide in the same country or group of countries through historical, political, economic, geographical or social causes, these become critical scenarios where convergence between terrorism and crime may radicalise and spread. It is important to remark that more than half of terrorist's attacks perpetrated in 2013 (last year for which complete recordings have been demanded) took place in only three countries, Iraq, Pakistan and Afghanistan(National Consortium for the Study of Terrorism and Responses to Terrorism 2014) all of them outstanding examples of the critical scenarios detailed before and in which

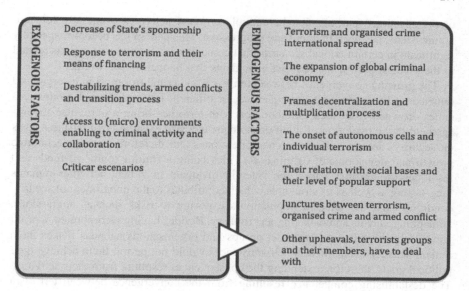

Illustration 2 Terrorist's implication factors

convergence between terrorism and crime reaches its highest rates (De la Corte 2013a, b). The western strip from Sahel[1] to Libya, so relevant for the security of all the countries settled in the Mediterranean's southern rim, nearby Spanish coastlines, are great examples of critical scenarios for the convergence between terrorism and crime, just as Syria, which is deeply related to Iraq's, Somalia and Nigeria scenarios (all included in the top ten list of terrorism more affected countries in 2013) (Illustration 2).

6 Consequences and Conclusions

According to the evidences based during neither the last years, the convergence between terrorism and crime is not a relationship that needs to be proved nor a simple anecdote. The main object of this paper, terrorist actors' (individuals, groups

---

[1]Researchers of the SahelinBoukara Houcine, Messaoud Fenouche, Touatit Lofti, Benhadj Karima, José María Blanco y Luis de la Corte Ibáñez, "Terrorismo y tráfico de drogas en África subsahariana", Madrid, Instituto Español de Estudios Estratégicos e Instituto Militar de Documentación, Evaluación y Prospectiva de Argelia. Project document from the Instituto español de Estudios Estratégicos, 13/3/2013. Available on: http://www.ieee.es/Galerias/fichero/docs_trabajo/2013/DIEEET01-2013_IEEE-IMDEP.pdf. Also on the Sahel Wolfram Larcher, "Organized Crime and Conflict in the Sahel-Sahara Region". Carnegie Endowment for International Peace, September 2012. Available on: http://www.carnegieendowment.org/files/sahel_sahara.pdf.

or organisations) participation in multiple criminal activities essentially for economic needs, has been frequent and progressive. This pattern has become especially significant in certain critical scenarios that coincide with countries or regions related to the greatest part of the terrorist's incidents that happen in the world.

The growing convergence between terrorism and crime (for most in its organised dimension) has numerous consequences for different levels. Criminal groups revenues drowned from their collaboration with terrorists bring a particular encouragement to the evolution of illicit economies, increase existing legal business costs on security, decrease efforts to create new ones and therefore restrains economic growth and development. Criminal activities being a fruitful source of funds and resources for terrorist actors, the latter involvement in violent activities remains long-lasting as well as its structures, which contributes to the impairment of security conditions and the affected population's exposure to risks such as aggressions, kidnapping, threats and racketing and robbery. Besides having incorporated, a profit gainful objective in some terrorist groups and organisations agendas makes them remain actively even when the circumstances would not permit them achieve their political goals, therefore extending the consequences resulting from their activities. The destabilising competence resulting from the convergence between terrorism and crime in countries in which this connection reaches its highest rates contributes to delve controllability and repressive inefficiency problems previously existing and creates a vicious dynamic in which both threatens coexist and feedback each other. On the other hand, the fact that such extreme convergence examples occurs outside from western countries does not avoid the cross-border prolongation of its effects. Quite the opposite, the intense noticed juncture in the so-called critical scenarios, some of which also operate as genuine terrorism epicentres with global spread pretensions (Afghanistan, Pakistan, Iraq, Syria, Sahel, etc.), and it also threatens international security and peace. In part because profits resulting from illicit ways gathered by terrorists' elements contribute to the persistence of armed conflicts that require the concern of the international community. And also because when financing and facilitating the flow of foreign radicals take place, the risk of spreading terrorism to other nation surrounding or distant (Spain and European countries included) through violent actions and campaigns headed by combatants returning from such scenarios. This project takes wilfully into account common and organised crime on minor scale for its considerations. Since the attacks of 11 March perpetrated in Madrid in 2004, which were financed by a range of small and humble criminal operations performed by a handful of common criminals, proved the dramatic consequences underrating such minor entities criminal ties can mean (Reinares 2014).

Concluding, the concern on terrorism and crime junctures probabilities is largely justified, especially when everything suggests these threaten will follow us for a long time. In this regard, the creation of coordination and cooperation mechanisms between security and intelligence departments with responsibilities on criminal and counter-terrorism researching opens new development opportunities.

References

Bovenkerk, F., & Chakra, B. A. (2004). Terrorism and Organized Crime, *Forum on Crime and Society, 4*, 1–2, 3–16.

Chernick, M. (2007). FARC-EP: Las Fuerzas Armadas Revolucionarias de Colombia-Ejército del Pueblo. In M. Heiberg, B. O'Leary, & J. Tirman (Eds.), *Terror, insurgency, and the state* (pp. 51–82). Philadelphia: University of Pennsylvania Press.

Cornell, S. E. (2006). The narcotics threat in greater central Asia: Fromm crime-terror Nexus to State infiltration. *China and Eurasia Forum Quarterly, 4*(1), 37–67.

Cronin, A. K. (2009). *How terrorism ends. Understanding the decline and demise terrorist campaigns* (pp. 146–152). Princeton, NJ: Princetown University Press.

De la Corte, L. (2013a). Criminalidad organizada y conflictos armados, *Ejército, 838*, 18–26.

De la Corte, L. (2013b). To what extent do global terrorism and organised criminality converge? General parameters and critical scenarios. *Journal of the Spanish Institute of Strategic Studies, 1*(1), 1–27. Available on: https://revista.ieee.es/index.php/ieee/article/view/41. Accessed on; June 6, 2016.

Díez, L. (2015).¿Cómo se combate el terrorismo yihadista en España?, *Cuarto poder*, 15/11/2015. Available on: http://www.cuartopoder.es/laespumadeldia/2015/11/15/como-se-combate-el-terrorismo-yihadista-en-espana/17466. Accessed: 3/6/2016.

Hoffman, F. (2007). *Conflict in the 21st century: The rise of hybrid wars*. Arlington: Potomac Institute for Policy Studies.

Hutchinson, S., & O'Malley, P. (2007). A crime-terror Nexus? Thinking on some of the links between terrorism and criminality. *Studies in Conflict & Terrorism, 30*, 1095–1107.

Makarenko, T. (2004). The crime-terror continuum: Tracing the interplay between transnational organised crime and terrorism. *Global Crime, 6*, 129–145.

Naim, M. (2006). *Ilícito. Como traficantes, contrabandistas y piratas están cambiando el mundo*, Debate, Barcelona.

National Consortium for the Study of Terrorism and Responses to Terrorism. (2014). Majority of 2013 terrorist attacks occurred in just a few countries, Maryland University, December 2014. Available on: http://www.start.umd.edu/news/majority-2013-terrorist-attacks-occurred-just-few-countries

Nesser, P. (2015). *Islamist terrorism in Europe. A history* (pp. 15–16). Oxford: Oxford University Press.

Reinares, F. (2014), *Matadlos. ¿Quién estuvo detrás del 11-M y porqué se atentó en España?* (pp. 97–126). Barcelona: Galaxia Gutemberg-Circulo de Lectores.

Sageman, M. (2007). *Leaderless Jihad: Terror networks in the twenty-first century*. Philadelphia: Pennsylvania University Press.

Shelley, L. I. (2015). *Dirty entanglements: Corruption, crime, and terrorism*. New York: Cambridge University Press.

The Soufan Group. (2016). The graveyard off the coast of Libya. *TSG IntelBrief*, June 2. Available on: http://soufangroup.com/tsg-intelbrief-the-graveyard-off-the-coast-of-libya/

United Nations Office on Drugs and Crime. (2010a), *Crime and instability. Case studies of transnational threats*, Vienna, 2010. Available on: http://www.unodc.org/documents/data-and-analysis/Studies/Crime_and_instability_2010_final_26march.pdf

United Nations Office on Drugs and Crime. (2010b). *The globalization of crime. A transnational organized crime threat assessment*, Vienna, 2010. Available on: http://www.unodc.org/unodc/en/data-and-analysis/tocta-2010.html

United Nations Security Council, Resolution 1373. (2001). p. 3 Available on: http://www.un.org/en/sc/ctc/specialmeetings/2012/docs/United%20Nations%20Security%20Council%20Resolution%201373%20(2001).pdf. Accessed: September 28, 2001.

Valiño, A., Buesa, M., & Baumert, T. (2010). The economics of terrorism: An overview of theory and applied studies. *The Economic Repercussions of Terrorism*, 3–37.

Index

A
Analysis, 122, 123, 125, 130

B
Big data, 103, 104–106, 108, 114
Black swans, 201, 204, 205, 217

C
Child trafficking, 223, 229, 230, 232, 235–239
Cocaine trafficking, 137, 154, 160, 161, 163
Community detection, 190
Copper theft, 167, 182–184, 191, 196
Crime, 119–123, 125–136, 263–270, 274–276, 278
Crime fighting, 103, 104, 106, 107, 109, 110, 112, 114
Crime indicators, 167, 171, 173, 174, 179, 182, 195
Crime relevant factors, 47, 49, 52, 53, 57
Crime threat detection, 48, 55
Criminal intelligence, 36
Criminals, 119, 122, 126, 133, 136

D
Drugs, 132, 135
Dystopias, 200, 203, 207–211

E
Early warning, 30, 38, 73–77, 79, 81–83, 89, 91, 97
Emerging topic detection, 170, 184
Enforcement, 119, 131
Environmental scanning, 47–49, 49–55, 66, 68
Environmental scanning methodology, 74, 75, 79, 80
Exploitation, 119, 132, 133
Extremism, 242–244, 247, 252, 254, 257

F
Factors, 119–121, 125–129, 131, 134, 135
Foreign fighters, 241, 245, 246, 250
Foresight, 73–75, 77, 83, 95, 99, 199, 201, 203, 205, 207, 208
Future, 137–141, 143, 145–151, 154, 157, 159, 165, 199–212, 217
Future crime, 73–75, 79

H
Horizon scanning, 75
Human trafficking, 233, 235, 237–239

I
Indicators, 25, 29, 32, 33, 36–40
Intelligence, 120–124, 136, 137, 141, 144, 148, 150, 154, 165
Intelligence-led policing, 223, 224
Investigation, 119, 120, 130
Investigation case, 223, 224, 236, 239

J
Jihadist propaganda, 167, 182, 185, 187, 189, 191, 192, 196

L
Lone-wolf, 241, 245, 250, 253, 257

M
Media and terrorism, 241, 247, 251–253, 255, 256
Methodologies of measuring organised crime, 25, 27, 32, 35, 40
Methodology, 141, 143, 147, 148

N
National security, 103, 104, 109
Nexus Terror-Crime, 261, 262

O

Open data, 47, 48
Open source intelligence, 65, 223–225
Organised crime, 3–13, 15–19, 25, 29–39, 47, 48, 51–53, 55, 60, 137, 139–141, 143, 147, 149, 151, 153, 159, 160, 165, 199, 200, 203, 207, 209, 211, 213, 214, 216, 217, 261–263, 265, 267, 270, 271, 273–275, 278
Organised crime concept, 3, 4, 6, 9, 18
Organised crime definition, 3, 4, 6, 8
Organised crime measurement, 3, 10, 19
OSINT, 179, 181

P

Predictive policing, 167–170, 173
Predictive policing weak signals, 73, 74, 90, 93, 97
Privacy, 103, 104, 107–115

Q

Quantitative data, 25, 31, 32

R

Radicalization, 241, 242, 244, 245, 247, 248, 250–254, 257
Radicalization drivers, 241–243, 247, 248
Radicalization indicators, 241, 244, 246, 257

S

Scenario, 120, 121, 130–132, 134, 135
Security, 120–122, 125, 130, 132, 134
Slavery, 133
Social media monitoring, 171–174, 181
Social network analysis, 168, 170, 171, 176
SOCMINT, 167, 179, 180
Strategic early warning, 48, 53, 55, 68

T

Terrorism, 241–244, 246, 247, 255, 257, 261–265, 267, 269–276, 278
Terrorist scenarios, 241, 242
Text mining, 167, 170, 171, 177, 179
Trafficking, 119, 130, 132–135
Trends, 121, 122, 125, 127, 129–131, 138, 139, 147, 149–160, 163, 165
Trust, 103, 104, 112, 113, 115

V

Visual analytics, 173, 175, 177, 179
VUCA, 137, 139, 140, 142

W

Weak signals, 36–39
Wild cards, 201–206, 208, 211

CPSIA information can be obtained
at www.ICGtesting.com
Printed in the USA
LVOW05*1513051217
558724LV00002B/25/P